The Open University

A Second Level Course

INTRODUCTION TO ENGINEERING MECHANICS

Unit 8
Vibrations and Dimensional Analysis

Unit 9
Examples in Machine Design

Prepared by the Course Team

THE OPEN UNIVERSITY PRESS

The Introduction to Engineering Mechanics Course Team

Authors

J. K. Cannell	(Engineering mechanics) Chairman
O. R. Fendrich	(Engineering mechanics)
G. S. Holister	(Engineering science)
P. J. Lucas	(Engineering mechanics)
V. Marples	(University of Warwick)
P. Minton	(Imperial College, University of London)
R. K. Pefley	(University of Santa Clara, California)
C. N. Reid	(Materials)
B. O. Shorthouse	(Staff Tutor)

Other members

A. J. Crilly	(BBC)
R. P. Dobson	(Course Assistant)
R. D. Harrison	(Educational Technology)
A. R. Sollars	(Staff Tutor)
S. J. Stickland	(Editor)
V. Woodhouse	(Technician)
P. R. V. Youngman	(Scientific Officer)

The Open University Press
Walton Hall, Milton Keynes

First published 1975

Designed by the Media Development Group of the Open University.

Printed in Great Britain by
Martin Cadbury, a specialized division of Santype International.
Worcester and London.

ISBN 0 335 02854 3

This text forms part of an Open University course. The complete list of units in the course appears at the end of this text.

For general availability of supporting material referred to in this text please write to the Director of Marketing, The Open University, PO Box 81, Milton Keynes, MK7 6AT.

Further information on Open University courses may be obtained from the Admissions Office, The Open University, PO Box 48, Milton Keynes, MK7 6AB.

1.1

Unit 8
Vibrations and Dimensional Analysis

CONTENTS

AIMS

The first part of this unit aims to extend the vibration studies introduced in Units 6/7 to the situation where an external periodic force excites and maintains vibration in a single degree-of-freedom dynamical system; this system has properties of mass, stiffness and damping. The unit discusses how such periodic forces arise, and also the behaviour and theory of resonant dynamical systems. The unit also aims to introduce various criticisms and limitations of the one degree-of-freedom model and outlines other means of modelling real systems.

The second part of the unit introduces a semi-empirical method of analysing problems, called dimensional analysis. This is based on a consideration of the basic dimensions of the physical quantities in the variables involved. It is often useful in tackling problems which are too complex to permit rigorous mathematical treatment. The solution so obtained is generally incomplete but it indicates how the final details may be experimentally determined. The aim of this part of the unit is to enable you to apply this method of analysis to appropriate problems and to enable you to see what would then be decided by experiment.

OBJECTIVES

After studying this unit you should be able:

1   To explain the difference between the analysis of free vibrations and that of forced vibrations.

2   To use the solution to the differential equation $m\ddot{x} + c\dot{x} + kx = F \sin \omega t$.

3   To calculate quantitative results concerning single degree-of-freedom forced, damped vibrations using graphs of vibratory response.

4   To know what is meant by the following terms: phase angle, resonance, temporal lag, and amplification factor.

5   To define, and explain the difference between, geometric, kinematic and dynamic similarity.

6   To explain how dimensional analysis can be used to set up a relationship between dimensionless groups of variables in specific instances, and to form some such dimensionless groupings or numbers yourself.

7   To explain how such a relationship can be used to dictate the nature of experimental work aimed at obtaining an explicit relationship.

8   To indicate some of the ways in which there might be difficulty or uncertainty in achieving a dynamically similar model using the results of dimensional analysis.

STUDY GUIDE

There is a television programme ('Vibrations') and a disc associated with this unit. You will also be able to do Experiment 6 on forced vibrations when you have read the first part of the unit.

You will not be examined on the material in the appendices to the unit, but you should be able to *use* the rule discussed in Appendix B.

If you are short of time, the sections on which you should particularly concentrate in Part I of this unit are Sections 3 to 3.4. The bulk of Part II, contained in Sections 7 and 8, is equally important.

Part I

Vibrations

INTRODUCTION

The first half of this unit continues the discussion and analysis of vibrations that we began in the preceding unit. As before, the vibrations which will concern us apply to solid body systems having properties of mass, stiffness and damping, or components that provide these properties. Most of the work that follows relates to *single* degree-of-freedom systems that are capable of vibrating, or at least responding to perturbations, in a cyclical, periodic manner. But whereas in Units 6/7 we concentrated entirely on freely vibrating systems (those in which a free and natural vibration of the system followed some initial disturbance), in this unit we will concentrate on systems subject to *forced* vibrations. Forced vibrations do not decay unless the cyclic force which causes them changes. Even then they rarely disappear unless the impressed force also disappears.

As you will see, the transient response of such a system before it settles down to its regular steady-state vibration is closely linked to its free vibration response, as examined in Units 6/7. When the impressed periodic force continues to act and the resulting vibration of the responding system reaches a steady state, the amplitude of that vibration depends partly on the amplitude of the cyclic force. However, this displacement may not be in phase with the force. It is the match between the frequency of the force which is exciting the oscillatory response, and the system's own properties of mass, stiffness and damping, that principally determines the scale of the consequent response.

Forced vibrations (including resonant ones which are the most obvious) occur all around us. The panel resonances in a motor car at low speeds; the rattling of house windows as a heavy lorry drives past; the slapping of the rigging and stays of a boat in a wind; the wobbling of an ill-balanced washing machine or the noise of a badly suspended refrigerator motor; all these are good examples of forced vibrating systems. Your home kit experiments on vibrations clearly contrast a forced vibration situation (Experiment 6) with a free vibration model (Experiment 5).

'The rattling of house windows as a heavy lorry drives past . . .'

One important branch of engineering practice deals almost exclusively with transient and forced vibrations in machines. Difficulties with machine vibration usually occur because machines generate their own periodically varying forces. These forces inevitably act on parts of the machine which have mass and stiffness, and produce 'forced vibration.' We should therefore look in detail at how a simplified mass–stiffness–damping system (Figure 1) responds to an externally impressed force.

Fortunately, in this simple case the approach to the problem is relatively straightforward. All we need to do is to express the exciting force as a function of time and to equate it to the dynamic forces involved in the oscillatory system. For instance if the exciting force can be expressed as some function of time, $F'(t)$, equation (31) of Units 6/7 becomes modified to

$$m\ddot{x} + c\dot{x} + kx = F'(t). \tag{1}$$

All we are doing here is applying Newton's second law and saying that the sum of the dynamic forces of the mass, damping and stiffness can be equated to the exciting force instead of to zero.

So, if we can analyse the exciting force sufficiently to express it as a function of time, it only remains to solve the new differential equation to find out how the system behaves. It can be shown (though the methods are beyond the scope of this course) that most forcing functions can be expressed in terms of sums of simple sinusoidal functions of different frequencies. We shall look at only the simplest of such possibilities where the exciting force $F'(t)$ is given by

$$F'(t) = F \sin \omega t. \tag{2}$$

In equation (2), $F'(t)$ is the value of the force at any instant, F is the amplitude (the maximum value of the force), and ω is the angular frequency of the force (so that its cyclic frequency is $\omega/2\pi$ Hz).

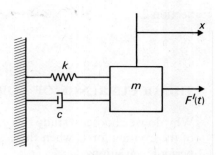

Figure 1   Representation of a one degree-of-freedom system with mass, stiffness and damping

THE OCCURRENCE OF FORCED VIBRATIONS

Why do we choose to study this particular case of a sinusoidal variation of the exciting force, when there must be an infinite number of possible periodic variations?

There are in fact good reasons for selecting it as we shall now show.

There are many cases of vibration excitation in which the forces are obviously periodic, and in which they are (perhaps less obviously) sinusoidal. In machines, such periodic forces can be caused by the dynamic forces from fans or propeller blades, by gear-tooth forces, by cutting-tool forces (in machine tools) and by mass unbalance. There are also other less apparent causes.

Mass unbalance is one of the most common sources of periodic excitation. If any rotor which is supported in a pair of bearings (Figure 2) has a centre of mass which does not coincide *exactly* with the centre line of the bearing, then when the rotor is rotated at speed, a centripetal force will be required which can be produced only by the reactions: that is, by the forces exerted by the rotor bearings. These forces must rotate as the shaft rotates. Let the forces be F_A and F_B.

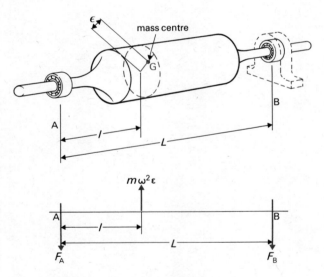

Figure 2   A rotor supported on bearings A *and* B

At every instant we must have

$$F_A + F_B = m\omega^2\epsilon$$

and by taking moments about the plane of rotation of the centre of mass G, we have

$$F_A \times l - F_B \times (L - l) = 0$$

since the rotor axis has no angular acceleration. Hence

$$F_B \times L = m\omega^2\epsilon l$$

where m is the mass of the rotor and G is the position of the centre of mass, a distance l from the left-hand bearing and at a radius ϵ from the bearing centre line.

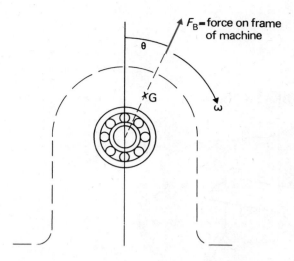

Figure 3

F_B is a force of magnitude $m\omega^2\epsilon l/L$ which *rotates* with angular velocity ω and always acts in a direction opposite to that of the rotor reactions to the centripetal force on the rotor. Thus at the instant shown in Figure 3, the machine which contains the rotor feels a vertical force of $F_B \cos\theta$ through the bearing at B and a horizontal force of $F_B \sin\theta$. The rotor is rotating at constant angular velocity ω and so the horizontal and vertical forces are $F_B \sin\omega t$ and $F_B \cos\omega t$ respectively, since $\theta = \omega t$. Similar forces are applied to the machine at the bearing A, where

$$F_A = m\omega^2\epsilon\left(\frac{L-l}{L}\right),$$

so there is sinusoidal excitation of the machine in both the vertical and the horizontal directions.

The forces transmitted and generated between pairs of meshing gear-teeth vary, not once every revolution, but once every time a new pair of gear-teeth come into mesh. Thus the periodic variation can be represented by expressions containing terms like $\sin n\omega t$ where n is the number of teeth on a gear-wheel rotating with angular velocity ω. The frequency $n\omega/2\pi$ is called the gear-tooth contact frequency and can often be readily identified as a discrete component of the noise generated by geared machinery. With the forces from fans and propellors there is a similar periodicity, where ω is the blade-passing frequency.

In metal machining processes there is periodicity in the cutting-tool contact frequency. In these cases the amplitude of the exciting force is not necessarily constant, nor does the force vary simply sinusoidally despite its true periodicity. Imagine a single-point cutter machining a semicircular groove in the edge of a metal plate (Figure 4(a)). The cutter will be subject to a force due to cutting of the metal only during every other half rotation. And if we assume that the force is constant during cutting, the graph of cutter force versus time is like that shown in Figure 4(b). Such a force function can be resolved into a sum of sine and/or cosine functions. In this case the function is

$$F'(t) = F/2 + (2F/\pi)\sin\omega t + (2F/3\pi)\sin 3\omega t + (2F/5\pi)\sin 5\omega t + \cdots \quad (3)$$

(we will not consider here how the various terms of this equation arise).

This function could be used in equation (1), giving further justification for learning to solve it with $F'(t)$ equal just to $F \sin\omega t$: the complete solution is made up of the sum of solutions obtained by substituting each of the terms on the right-hand side of equation (3) separately into equation (1). Most real machining processes use multi-point tools and are

11

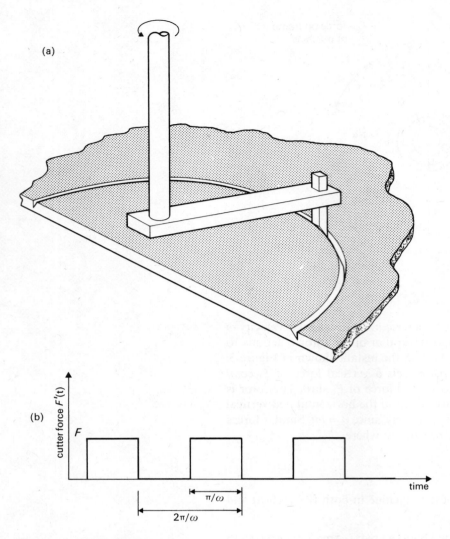

(a)

(b)

cutter force $F'(t)$

F

time

π/ω

$2\pi/\omega$

Figure 4   A single-point cutter

more complex than this one but even these can be analysed into a sum of a series of sine and cosine terms of integer multiples of ωt.

The example of the single point cutter also reveals the varied nature of vibration. Because the point of the tool is eccentric with respect to the end of the tool shaft (Figure 4(a)) there is both a torque and a bending moment applied to the shaft. Thus the shaft could be subject to both torsional oscillation and vibration in bending, in addition to any vibration that is generated in the machine itself by the fluctuating forces transmitted through the shaft bearing. Whether or not the amplitudes of any of these types of vibration would be significant depends on the appropriate mass, stiffness and damping values. This follows because, just as in the case of free vibration, we could interpret equation (36) in Units 6/7 to be for vertical, horizontal or torsional motion, so we can interpret equation (1) to be for the vertical, horizontal or torsional movement of the same machine, depending solely on the values which we insert for the parameters.

The significance of this general description of vibration will become clearer with increasing experience. Let us now set about solving the differential equation.

Section 3 gives a simple treatment of the performance of a single degree-of-freedom spring–mass–damper system under the action of a sinusoidal force. You should not try to remember the mathematics involved, but you should be able to follow the logical steps in it. You should understand the physical significance of the result and be able to use it.

SOLUTION OF THE FORCED DAMPED VIBRATION EQUATION

(Appendix A shows an alternative method of solving this equation: you should read this after working through Section 3 if you have time.)

The equation of motion for simple *forced* vibration of a damped single degree-of-freedom system is

$$m\ddot{x} + c\dot{x} + kx = F \sin \omega t \tag{4}$$

where ω is *not* to be confused with the (undamped) natural angular frequency of the system, ω_n. The natural frequency ω_n is a function of the parameters of the system, m and k, while ω is determined entirely by the means of excitation and so is entirely independent of ω_n.

In the case of free vibration we saw how we could get a solution for the second order differential equations in terms of the sum of a sine and a cosine term. Let us try this for the forced vibration case. Since the system will eventually settle down to a motion with the same frequency as the external force, let us assume that we have a solution of the form

$$x = D \sin \omega t + E \cos \omega t \tag{5}$$

where D and E are constants. Differentiating gives

$$\dot{x} = D\omega \cos \omega t - E\omega \sin \omega t$$

and

$$\ddot{x} = -D\omega^2 \sin \omega t - E\omega^2 \cos \omega t.$$

So, substituting for x, $\dot{x}$ and $\ddot{x}$ in the left-hand side of equation (4),

$$\begin{aligned}
m\ddot{x} + c\dot{x} + kx &= -mD\omega^2 \sin \omega t - mE\omega^2 \cos \omega t + cD\omega \cos \omega t \\
&\quad - cE\omega \sin \omega t + kD \sin \omega t + kE \cos \omega t \\
&= [D(k - m\omega^2) - Ec\omega] \sin \omega t \\
&\quad + [E(k - m\omega^2) + Dc\omega] \cos \omega t
\end{aligned}$$

and this must be equal to $F \sin \omega t$ if the assumed solution is going to satisfy the equation. Therefore, since we want a solution which will be valid for all values of t, the coefficient of $\sin \omega t$ must be equal to F and that of $\cos \omega t$ must be equal to zero.

Hence

$$D(k - m\omega^2) - Ec\omega = F$$

and

$$E(k - m\omega^2) + Dc\omega = 0.$$

From the second of these equations

$$E = -\frac{Dc\omega}{(k - m\omega^2)}$$

and substituting this expression for E in the first of the equations gives

$$D\left[(k - m\omega^2) + \frac{c^2\omega^2}{(k - m\omega^2)}\right] = F$$

$$D = \frac{F(k - m\omega^2)}{(k - m\omega^2)^2 + c^2\omega^2}.$$

13

Therefore

$$E = \frac{-Fc\omega}{(k-m\omega^2)^2 + c^2\omega^2}.$$

If we now substitute these expressions for D and E in equation (5) and do a little rearranging, we have a solution of the differential equation (4):

$$x = \frac{F}{(k-m\omega^2)^2 + (c\omega)^2}[(k-m\omega^2)\sin\omega t - c\omega\cos\omega t]. \qquad (6)$$

This can alternatively be written as

$$x = \frac{F}{\sqrt{[(k-m\omega^2)^2 + (c\omega)^2]}}\sin(\omega t + \phi) \qquad (7)$$

where

$$\tan\phi = -\frac{c\omega}{k-m\omega^2}.$$

This last step can be explained if we write

$$(k-m\omega^2)\sin\omega t - c\omega\cos\omega t$$

in the form

$$A\sin\omega t + B\cos\omega t$$

where $A = (k-m\omega^2)$ and $B = -c\omega$. We can put

$$A\sin\omega t + B\cos\omega t = C\sin(\omega t + \phi)$$

provided that C and ϕ are the right functions of A and B.

These functions can be found by expanding the expression $C\sin(\omega t + \phi)$:

$$C\sin(\omega t + \phi) = C(\sin\omega t\cos\phi + \cos\omega t\sin\phi).$$

In order to make this equal to $(A\sin\omega t + B\cos\omega t)$, we must have

$$C\cos\phi = A \qquad \text{and} \qquad C\sin\phi = B.$$

Squaring and adding these equations we get

$$C^2(\cos^2\phi + \sin^2\phi) = A^2 + B^2$$

and since $\cos^2\phi + \sin^2\phi = 1$,

$$C = \sqrt{(A^2 + B^2)}.$$

Dividing the second equation by the first, we have,

$$\frac{\sin\phi}{\cos\phi} = \frac{B}{A} = \tan\phi.$$

Therefore, we have

$$A\sin\omega t + B\cos\omega t = [\sqrt{(A^2 + B^2)}]\sin\left(\omega t + \tan^{-1}\frac{B}{A}\right).$$

(Alternatively, this could have been obtained graphically from Figure 5.)

Thus we obtain the expression for $\tan\phi$:

$$\tan\phi = \frac{-c\omega}{k-m\omega^2}.$$

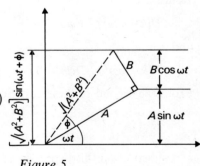

Figure 5

The angle ϕ is called the *phase angle*; it is the angle separating the force F' from the consequent displacement function x. We will discuss it more carefully later in the unit.

phase angle

Equations (6) and (7) represent a sinusoidal motion with a frequency *equal to that of the exciting force.*

Although equation (7) has been derived as a solution for x in equation (4), it is not a complete solution. It is only a partial solution, called the *particular integral*, because it refers to the particular type of forcing function which appears on the right-hand side of equation (1). If we were to add to this particular integral the expression for x in equation (42) of Units 6/7 and then determine $m\ddot{x} + c\dot{x} + kx$ we would *still* obtain $F \sin \omega t$ because $m\ddot{x} + c\dot{x} + kx$ for the values of x in equation (42) is zero.

particular integral

Therefore the expression

$$x = e^{-\alpha t}(A \sin pt + B \cos pt) + \frac{F}{\sqrt{[(k - m\omega^2)^2 + (c\omega)^2]}} \sin(\omega t + \phi) \quad (8)$$

is a more general solution to the differential equation (4) than is the solution given by equation (7).

Although we will not prove it here, it is the complete solution. The contribution from equation (42) in Units 6/7 is called the *complementary function* because it complements the particular integral.

complementary function

The first part of the general solution relates to the free vibration, often called the *transient vibration* because it is present only temporarily. The second part, which is directly dependent on both the amplitude and frequency of the exciting force, is referred to as the *steady-state vibration* since the amplitude of the sinusoidal oscillation which it represents does not vary with time. It is important to appreciate that the frequency of the free, transient vibration p in the first part of the solution and the frequency of the steady-state vibration ω in the second part are not necessarily the same. Sometimes by coincidence they may be very close but equally, either could be greater than the other.

transient vibration

steady-state vibration

Note again that ω is the angular frequency of the forcing function which causes this forced vibration; ω_n is the natural *undamped* (free vibration) frequency of the system, where $\omega_n = \sqrt{(k/m)}$; and p is the natural *damped* (free vibration) frequency of the system, where

$$p = \sqrt{\left(\frac{k}{m} - \frac{c^2}{4m^2}\right)}.$$

Hence ω_n and p are properties of the system which is forced to vibrate at a frequency of $\omega \, \text{rad s}^{-1}$.

Thus the second part of equation (8), the steady-state vibration, describes the whole situation in the case of the steady forced vibration of a single degree-of-freedom mass–spring–damper system, but only when the starting, transient vibration has had time to disappear.

SAQ 1

SAQ 1

A dynamic system which can be adequately modelled by equations (4) and (7) has the following properties: $m = 0.1 \, \text{kg}$, $k = 15 \, \text{N m}^{-1}$, and $F = 0.7 \, \text{N}$. If the steady-state amplitude at a forcing frequency equal to the undamped natural frequency of the system is not to exceed 20 mm, calculate the least viscous damping coefficient required. What will be the corresponding phase angle between the applied force and the motion of the mass?

3.1 Steady-state vibration

In this section we consider the steady-state part of a forced vibration.

A great deal of the theory of vibration is based on ignoring the first (transient) term of the complete solution of the differential equation of forced vibration. This is done on the assumption that the motion which it represents disappears completely after a short space of time. Attention is then concentrated on the second term. There is some merit in separating the two effects at this stage because while the amplitude of the transient vibration is principally a function of time, the steady-state vibration amplitude is principally a function of excitation frequency and not a function of time at all.

It is therefore unnecessarily complex to study both transient and steady state terms simultaneously, and here we will separate out the two effects in order to explain the significance of the steady-state term. However, in doing so we must remember that some real vibration states are actually more like transient than steady-state situations. By concentrating on the second term it is all too easy to think of all vibration states as being steady states and to attempt to model them without including the complementary function part of the total solution.

With this word of caution let us look at the steady-state part of the solution. It is periodic in ω, the amplitude being a function, not of time, but of excitation frequency $\omega/2\pi$. It is also a function of the parameters of the excited system. To understand the interdependence of the several variables, it is convenient to express the amplitude of vibration X as a multiple of the so-called static deflection, F/k. (This deflection is simply the static value of x which would result from the deflection of the spring under the action of a constant force of magnitude F when $\omega = 0$.) We have

$$\frac{X}{F/k} = \frac{1}{\sqrt{\left[\left(1 - \frac{m\omega^2}{k}\right)^2 + \left(\frac{c\omega}{k}\right)^2\right]}} = \frac{1}{\sqrt{\left[\left(1 - \frac{\omega^2}{\omega_n^2}\right)^2 + \left(\frac{2\alpha\omega}{\omega_n^2}\right)^2\right]}}. \quad (7a)$$

Remember from Units 6/7 that $c = 2\alpha m$ where α is the *decay factor*. Thus

$$\frac{c\omega}{k} = \frac{2\alpha m\omega}{k},$$

and since $k/m = \omega_n^2$ (where ω_n is the natural undamped angular frequency),

$$\frac{c\omega}{k} = \frac{2\alpha\omega}{\omega_n^2} = 2 \times \frac{\omega}{\omega_n} \times \frac{\alpha}{\omega_n}$$

where α/ω_n is the damping ratio and is equal to c/c_c.

The significance of this expression, and the interdependence of the variables involved, can now be shown by plotting the dimensionless *amplification factor* Xk/F against the dimensionless *frequency ratio* ω/ω_n, for different values of damping ratio (Figure 6(a)). By 'dimensionless,' we mean that the values of the resulting variables or numbers do not depend on the set of units used for measuring x, ω, F etc. This concept is explained more fully in Part II of this unit.

amplification factor
frequency ratio

Before looking at the graph in detail, it is worth observing that we can rapidly get an overall impression of the interdependence by considering just three different orders of magnitude of the excitation frequency in equation (7a).

(a) If ω is small with respect to ω_n (written $\omega \ll \omega_n$). Then

$$\left(\frac{\omega}{\omega_n}\right) \ll 1$$

and, since α/ω_n will probably be less than 1.0, we also have

$$\frac{2\alpha\omega}{\omega_n^2} \ll 1.$$

16

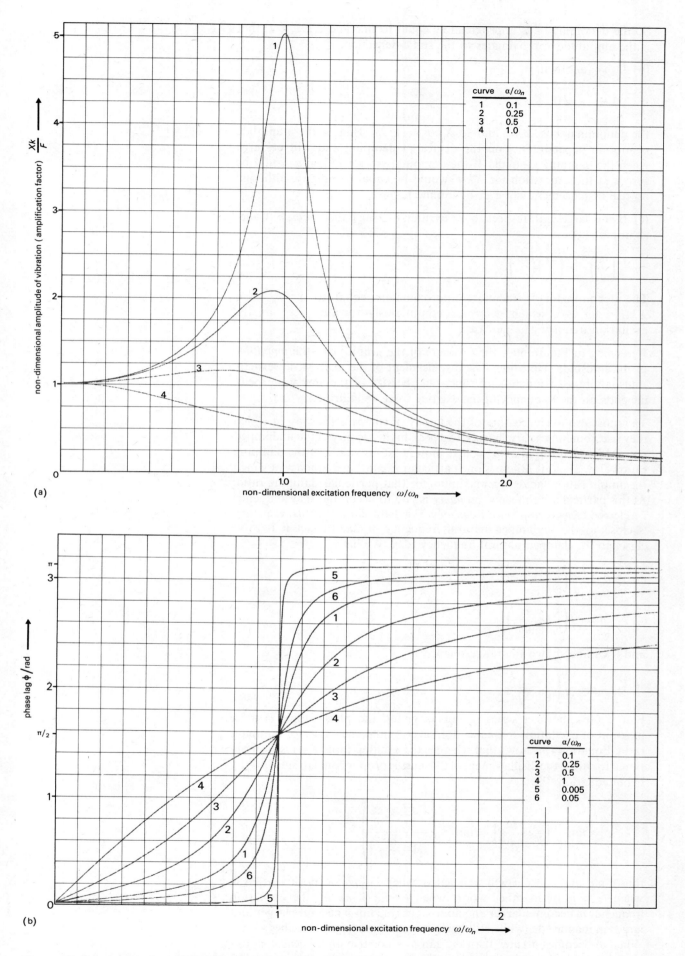

Figure 6 (*a*) *Amplitude and* (*b*) *phase angle as a function of excitation for different values of damping ratio*

So the denominator is approximately equal to one and $Xk/F \approx 1$, that is the amplitude approximates to the static deflection.

(b) If $\omega = \omega_n$, then

$$1 - \frac{\omega^2}{\omega_n^2} = 0,$$

the denominator is $2\alpha/\omega_n$ and so $Xk/F = \omega_n/2\alpha$. That is, the amplitude of motion depends only on the degree of damping and will be large when the damping is small. (If the damping was zero, then α would be zero and the expression for Xk/F would be equal to infinity, although no real displacement can ever be infinite.)

(c) If ω is large with respect to ω_n (written $\omega \gg \omega_n$), then $\omega/\omega_n \gg 1$ and, as in part (a),

$$\left(\frac{\omega}{\omega_n}\right)^2 \gg \left(\frac{2\alpha}{\omega_n}\right)\left(\frac{\omega}{\omega_n}\right).$$

The denominator is therefore approximately equal to ω^2/ω_n^2 and $Xk/F \approx (\omega_n/\omega)^2$, which decreases continuously towards zero as the excitation frequency ω increases.

Returning to Figure 6(a), the ordinate of the graph gives the *amplitude* of the sinusoidal vibration. The displacement of the mass varies exactly sinusoidally as a function of time, but the size of the sine wave peak depends on the frequency of the exciting force as indicated.

The most prominent feature of the family of curves is the way in which they rise, sometimes very sharply, to a peak in the vicinity of the undamped natural frequency f_n at $\omega/\omega_n = 1$. It shows that for values of the damping ratio below about 0.7 there is one particular excitation frequency at which the amplification factor is a maximum for that particular damping ratio. At this particular frequency, *resonance* is said to occur and that frequency is referred to as a *resonance frequency*. For light damping this frequency is close to the undamped natural frequency $\omega_n/2\pi$. In fact it is even closer to the damped natural frequency $p/2\pi$. We have

resonance
resonance frequency

$$\omega_n = \sqrt{\left(\frac{k}{m}\right)},$$

$$p = \sqrt{\left(\frac{k}{m} - \frac{c^2}{4m^2}\right)}$$

and it can be shown that at resonance

$$\omega_{\text{res}} = \sqrt{\left(\frac{k}{m} - \frac{c^2}{2m^2}\right)}.$$

Figure 6(b) shows a series of curves giving the phase angle ϕ as a function of excitation frequency for several different values of damping ratio. Equation (7) shows that whereas the exciting force $F'(t)$ is proportional to $\sin \omega t$, the displacement of the mass is proportional to $\sin(\omega t + \phi)$. From equation (7),

$$\phi = \tan^{-1}\left(\frac{-c\omega}{k - m\omega^2}\right) = \tan^{-1}\left(\frac{-\dfrac{2\alpha}{\omega_n} \times \dfrac{\omega}{\omega_n}}{1 - \dfrac{\omega^2}{\omega_n^2}}\right),$$

thus $\tan \phi$ and hence ϕ will be negative when $k - m\omega^2$ is positive, that is when $\omega_n > \omega$ or in other words when $\omega < \sqrt{(k/m)}$. As the excitation frequency ω becomes larger and approaches ω_n, $\tan \phi$ becomes larger and larger in magnitude (while still being negative) and ϕ approaches $-90°$. When ω becomes greater than ω_n, $\tan \phi$ is positive and, because ϕ has already passed through the value $-90°$, its value now lies between $-90°$

and $-180°$. When the excitation frequency becomes very large with respect to the natural frequency of the system, ϕ approaches $-180°$.

These changes are shown in Figure 6(b). The ordinates of all the curves are zero at zero excitation frequency, they again coincide at the value of $90°$, and they approach $180°$ as ω becomes very large. The effect of damping on this phase angle ϕ (by which displacement x lags behind the excitation force F) is only to change the rate at which ϕ changes between these values of 0, $90°$ and $180°$.

3.2 Phase angle

The description of the way in which ϕ changes does not help to explain what it is or give us any mental picture of its meaning. This is a serious shortcoming because phase is one of the important parameters in vibration, but it is one which is often incompletely understood.

Although a phase relationship is always expressed as an angular measure it is really indicative of a time relationship. In a vibratory system an alternating force will give rise to an alternating displacement. At low frequencies of alternation the peak displacement will occur at the same time as the force peak. But at higher frequencies the force will be changing so rapidly that the displacement can no longer keep pace with it but will lag behind it, the rate of alternation still being the same. You can get a good idea of lag by lightly holding one end of a ruler between your thumb and forefinger, with the ruler hanging vertically down. If you now move your hand from side to side in the plane of the ruler, you will find that at low frequencies the ruler moves *in phase* with your hand. At high frequencies the ruler will reach its extreme left-hand position approximately at the instant when your hand is at the extreme right-hand end of its travel, and vice versa. The motion of the ruler is then almost exactly *phase opposed* to that of your hand.

We could describe the lag of the vibration with respect to the force as, for example, 2 ms (milliseconds) at 50 Hz. But 20 μs (microseconds) at 5000 Hz represents the same degree of lag (work it out in relation to the periodic time in each case), so it is convenient to express the lag in a way which is not dependent on frequency. This is achieved by dividing the actual time lag, the *temporal lag*, by the time for one period of the vibration. We can think of this non-dimensional lag in radians or degrees, and it is known as the *phase angle*. (In more complex systems, the lag can just as easily be a lead, if we consider a lag greater than $180°$ as a lead.) It is this phase angle which has been symbolized as ϕ in equation (7) and which is plotted in Figure 6(b).

temporal lag

Figure 7 gives a more graphic portrayal of the meaning of phase angle. It is intended to portray the instantaneous positions of a spring-supported body which is constrained so that it can oscillate in the vertical direction only as indicated by the rollers (Figure 7(a)). Each of Figures 7(b)–(f) refers to a particular frequency ratio (the ratio of excitation frequency to the natural undamped frequency of the system) and shows a graph of the displacement of the mass plotted against time. (The mass and the spring are sketched in Figure 7(d) but have been omitted from the other figures). Figure 7(g) portrays the variation of excitation force level with time. The arrows show the magnitude and the direction at the corresponding points in the cycle; the first arrow (at the extreme left of Figure 7(g)) represents the maximum upward force. The corresponding maximum displacements in Figures 7(b)–(f) are indicated by a cross on the curves.

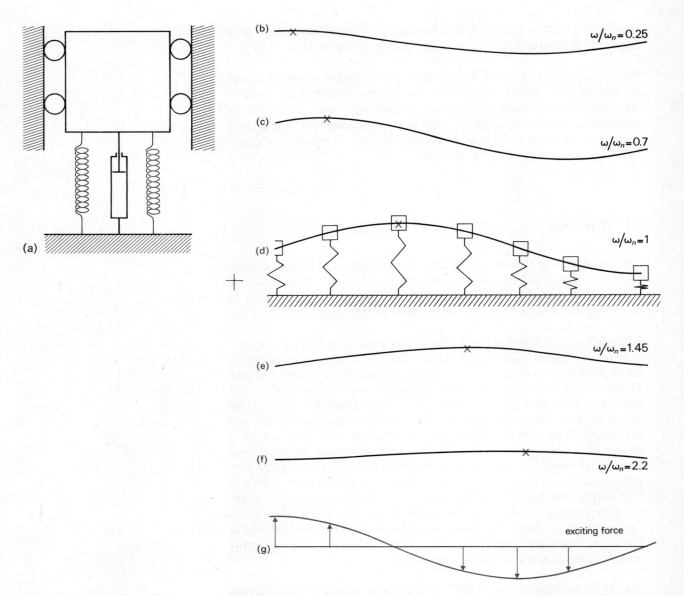

Figure 7   Vibration of a spring-supported body under the influence of an exciting force, showing the phase lag

The graphs in Figures 7(b)–(f) are sine waves, as is the force curve of Figure 7(g), but they are shifted to the right compared to the force curve. It is this shift which the phase angle measures. As the shift is to the right, indicating that the displacement reaches a peak level at a time later than that at which the force reaches a peak level, the phase of the displacement with respect to the force is lagging.

Both this sequence of diagrams and the more succinct and universal graphs of Figure 6 show that at low frequencies and low damping there is little or no temporal lag, but that as the frequency and/or damping increase so the lag increases. When the force has a frequency equal to the undamped natural frequency of the vibrating system the phase angle is 90° for any degree of damping. At frequencies higher than the natural frequency the phase lag is greater than 90°, and the amplitude of vibration is smaller than it is at the natural frequency. At still higher frequencies, irrespective of the degree of damping, the amplitude of vibration becomes nearer and nearer to zero, and the phase lag approaches 180°.

Exercise A

A machine is modelled as a mass of 150 kg supported by a device with a stiffness of $1.65\,\mathrm{MN\,m^{-1}}$ and a coefficient of damping of $3.14\,\mathrm{kN\,s\,m^{-1}}$.

It is excited into vibration by a sinusoidally varying force of amplitude 625 N and frequency 30 Hz. What will be the amplitude of vibration and what is the time lag of the displacement behind the force?

The undamped natural angular frequency of the system is

$$\omega_n = \sqrt{\frac{k}{m}} = \sqrt{\frac{1.65 \times 10^6}{150}} = 104.9 \,\text{rad s}^{-1}.$$

Therefore,

$$\frac{\omega}{\omega_n} = \frac{2\pi \times 30}{104.9} = 1.80.$$

Thus the damping ratio is

$$\frac{\alpha}{\omega_n} = \frac{c/2m}{\omega_n} = \frac{3.14 \times 10^3}{2 \times 150 \times 104.9} = 0.1.$$

Therefore, either by substituting in equation (7) or (7a), or by reading off the ordinates of curve 1 of Figure 6(a), we can obtain the amplification factor. (This is the ratio of the steady-state vibration amplitude to the static deflection due to a steady force, the magnitude of the steady force being equal to the amplitude F of the sinusoidal exciting force causing the vibration.) It is given by

$$\frac{Xk}{F} = 0.44.$$

Thus the amplitude of vibration

$$X = \frac{0.44 \times 625}{1.65 \times 10^6} = 0.17 \,\text{mm}.$$

From Figure 6(b), for the same frequency ratio,

$$\phi = 2.89 \,\text{rad}$$

which represents $2.89/2\pi$ of the cycle time. The cycle time is $2\pi/\omega$ seconds and so this phase lag represents

$$\frac{2.89}{\omega} = 15.3 \,\text{ms}.$$

Alternatively, as the system vibrates, the force 'vector' rotates at a rate of ω rad s^{-1}. Thus 2.89 rad displacement lag represents

$$\frac{2.89}{\omega} = \frac{2.89}{60\pi} = 0.0153 \,\text{s} = 15.3 \,\text{ms}.$$

SAQ 2

SAQ 2

In the machine of the previous exercise, it is decided that for the same damping ratio, an amplification factor of vibration Xk/F of 1.2 can be tolerated at frequencies above resonance. To what value can the stiffness be increased and what will then be the new damping coefficient and the new phase angle? (Use Figures 6(a) and 6(b).)

Exercise B

Show that for steady-state forced vibrations as described by equation (7), the ratio kX/F cannot be greater than unity at any frequency unless the damping ratio is smaller than $1/\sqrt{2}$.

We have (equation (7a)),

$$\frac{kX}{F} = \frac{1}{\sqrt{\left[\left(1 - \frac{\omega^2}{\omega_n^2}\right)^2 + \left(\frac{2\alpha\omega}{\omega_n^2}\right)^2\right]}}.$$

Let the frequency ratio $\omega/\omega_n = \lambda$ (for conciseness): note that λ cannot be negative, that is $\lambda \geqslant 0$.

(a) Suppose the damping ratio $\alpha/\omega_n = 1/\sqrt{2}$. Then

$$\frac{kX}{F} = \frac{1}{\sqrt{\left[(1-\lambda^2)^2 + \left(\frac{2\lambda}{\sqrt{2}}\right)^2\right]}} = \frac{1}{\sqrt{[(1-\lambda^2)^2 + 2\lambda^2]}}$$

$$= \frac{1}{\sqrt{[(1-2\lambda^2+\lambda^4+2\lambda^2)]}}.$$

Thus in this case the ratio will equal unity for $\lambda = 0$ and will be smaller than unity for $\lambda > 1$.

(b) Now suppose $\alpha/\omega_n > 1/\sqrt{2}$. We put

$$\frac{\alpha}{\omega_n} = \frac{1+a}{\sqrt{2}}$$

where a is positive. Hence

$$\left(\frac{2\alpha\omega}{\omega_n^2}\right)^2 = \left[\frac{2\lambda}{\sqrt{2}}(1+a)\right]^2 = 2\lambda^2(1+a)^2$$

and

$$\frac{kX}{F} = \frac{1}{\sqrt{[(1-\lambda^2)^2 + 2\lambda^2(1+a)^2]}}.$$

Thus in this case when $\lambda = 0$, $kX/F = 1$ and for all values of $\lambda > 0$ (for example $\lambda = \frac{1}{4}$, $\lambda = \frac{1}{2}$, $\lambda = 1$), $kX/F < 1$.

(c) Suppose $\alpha/\omega_n < 1/\sqrt{2}$. We put

$$\frac{\alpha}{\omega_n} = \frac{1-a}{\sqrt{2}}$$

where a is positive. Then,

$$\frac{kX}{F} = \frac{1}{\sqrt{[(1-\lambda^2)^2 + 2\lambda^2(1-a)^2]}}.$$

Thus now when $\lambda = 0$, $kX/F = 1$ but when $\lambda > 0$,

$$\frac{kX}{F} = \frac{1}{\sqrt{[1-2\lambda^2+\lambda^4+2\lambda^2(1-a)^2]}}$$

and this can be greater than unity. For example putting $\lambda = 1$ and $a = \frac{1}{2}$, gives

$$\frac{kX}{F} = \frac{1}{\sqrt{(1-2+1+\frac{1}{2})}} = \sqrt{2} \simeq 1.414.$$

Thus we have shown that resonance can occur only when $\alpha/\omega_n < 1/\sqrt{2}$.

SAQ 3

A dynamic system can be modelled by Figure 1. It has a resonant circular frequency of $30\,\text{rad s}^{-1}$ and a damping ratio of 0.25. The exciting force (in newtons) is $F'(t) = (2\sin \omega t)$, and $m = 0.1\,\text{kg}$.

Using the graphs in Figures 6(a) and 6(b), estimate for steady-state vibration:

(a) the maximum amplitude of motion of the mass;

(b) the range of values of ω which will keep the value of kX/F greater than 1.1;

(c) the phase angle at the lowest and highest frequencies within this range.

3.3 Reducing vibration – an application of the steady-state vibration theory

In this section we look at one way in which the family of curves in Figure 6(a) can be used in practice.

Consider a physical system which can be adequately represented by the model shown in Figure 1. When the mass is acted on by a sinusoidally varying force of constant amplitude, its response is represented by one of a family of graphs like those in Figure 6. Suppose now that it is necessary to reduce the motion of the mass as far as possible.

If we cannot alter the force, one possibility, at least in theory, is to increase the amount of damping. However, this would be really effective only if the original value of damping was small and if the frequency of the applied force was near to the undamped natural frequency of the system. For example if $\omega/\omega_n = 1.1$ and the damping ratio α/ω_n is 0.1, curve 1 of Figure 6(a) shows that the amplification ratio is about 3.3. If, by the addition of external devices, the damping ratio was increased to $\alpha/\omega_n = 0.25$ then at the same frequency ratio the amplification ratio would be reduced to about 1.7 (curve 2) which is a considerable reduction. On the other hand, if the same exercise were to be performed for a frequency ratio of 1.36, the amplification ratios before and after addition of the extra damping would be about 1.12 and 0.92 respectively, which gives a very small reduction in amplitude for a very large increase in damping ratio.

There is, however, another possibility for reducing the amplitude of vibration. This is to reduce the undamped natural frequency so as to raise the frequency ratio ω/ω_n, and so shift the characteristic point on the graph. This can be a simple and effective technique. For example, suppose a delicate machine on a heavy base resting on studded rubber sheet on a solid floor generates an exciting force at 12 Hz. It has a value for α/ω_n of 0.25 and a natural undamped frequency f_n of 8.8 Hz. The ratio ω/ω_n is thus given by

$$\frac{\omega}{\omega_n} = \frac{12}{8.88} = 1.36,$$

and hence from Figure 6(a), the amplification ratio is 1.12. However, if the base is supported on six rubber-in-shear mounts, which reduces the stiffness of the supports, the natural frequency of the system is reduced. Suppose it falls to 5.7 Hz. The value of ω/ω_n is then increased from 1.36 to 2.11, the characteristic point moves down the right-hand side of the curve of Figure 6(a) and the amplification ratio falls to 0.3.

This reduction of the support natural frequency is usually a far more reliable and effective way of reducing amplitudes of vibration than trying to increase the damping.

However, it is not always straightforward. Since $\omega_n = \sqrt{(k/m)}$, it is tempting to believe that the natural frequency can easily be depressed either by reducing the stiffness k (as in the example above) or by increasing the mass m (for example, by increasing the weight of the machine plinth). Unfortunately, there are practical limitations which must be taken into account.

The weight of the machine is supported statically on the mount, so that the static deflection of the mount d is given by $mg = kd$ and therefore $\sqrt{(k/m)} = \sqrt{(g/d)}$. Thus to reduce ω_n, the deflection of the suspension must be increased. Whether this is achieved by alteration in stiffness or mass is immaterial as far as the steady-state vibration is concerned. But if the mass is made too large the mounts become excessively expensive. If the stiffness is reduced, there is a possibility that the supporting points would become so flexible in relation to the load applied to them that the system would vibrate in a much more complex manner. It could therefore no longer be considered to have a single degree of freedom. A larger mass does have the advantage that if the system is disturbed by a sudden load or shock which injects a quantity of energy, the resulting additional motion of the system will be smaller than it otherwise would be.

As we saw at the beginning of this section, increasing the damping is not always a good method for reducing the amplitude. It does however increase the rate at which the effects of both shock and other transient disturbances can be dissipated.

Exercise

A machine of mass 1200 kg rests on anti-vibration mounts and is excited by a sinusoidally varying force of frequency 10 Hz. The anti-vibration mounts have a combined stiffness of $1.85\,MN\,m^{-1}$ and a combined damping coefficient of $14.0\,kN\,s\,m^{-1}$. Which of the following will lead to the smaller increase in vibration amplitude: adding steel springs to the support to increase the combined stiffness by 80%, or increasing the number of anti-vibration mounts in order to double both the combined stiffness *and* the combined damping coefficient?

The actual value of excitation force is irrelevant to the solution of the problem since the question requires a simple comparison of two states between which, it is presumed, the force remains the same.

Initially

$$\frac{\omega}{\omega_n} = \frac{2\pi \times 10}{\sqrt{(1.85 \times 10^6/1200)}} = 1.60$$

and

$$\frac{\alpha}{\omega_n} = \frac{c}{2m} \Bigg/ \sqrt{\frac{k}{m}} = \frac{c}{2\sqrt{(mk)}} = \frac{14.0 \times 10^3}{2\sqrt{(1200 \times 1.85 \times 10^6)}} = 0.15.$$

Therefore, by interpolating in Figure 6(a) or by substituting in equation (7a),

$$\frac{Xk}{F} = \frac{1}{\sqrt{\left[\left(1 - \frac{\omega^2}{\omega_n^2}\right)^2 + \left(\frac{2\alpha}{\omega_n} \times \frac{\omega}{\omega_n}\right)^2\right]}} = 0.613.$$

When steel springs are added, the stiffness is increased to (1.8×1.85) $MN\,m^{-1}$. Then ω/ω_n becomes 1.19 and α/ω_n becomes 0.112, thus Xk/F is increased to 2.02.

When stiffness and damping are both doubled, the stiffness is increased to $(2 \times 1.85)\,MN\,m^{-1}$ and the damping coefficient is increased to (2×14.0) $kN\,s\,m^{-1}$. Therefore ω/ω_n becomes 1.13 and α/ω_n is increased to 0.212, thus Xk/F becomes 1.81.

That is, the latter is the more attractive alternative.

3.4 Excitation by rotating unbalance

A common source of periodic exciting force in rotating machinery is mass unbalance. If the centre of mass m_e of a rotating component is at a distance ϵ from the centre line of the axis of rotation, a reaction to the centripetal force of magnitude $m_e\omega^2\epsilon$ will be transmitted to the supporting structure. We can think of this force as being a vector rotating at angular velocity ω (Figure 8). If we take a Cartesian (x, y) co-ordinate system, the vector can be resolved into orthogonal components F_x and F_y where

$$F_x = m_e\omega^2\epsilon \cos \theta,$$
$$F_y = m_e\omega^2\epsilon \sin \theta.$$

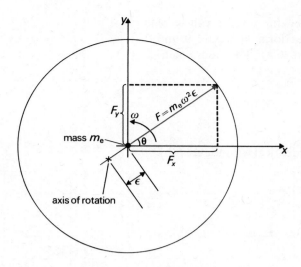

Figure 8   Force on supporting structure due to mass unbalance

If the angular velocity is constant we can write $\omega = \theta/t$, as long as we measure t from the instant when the vector F was directed wholly along the x axis. So we have $\theta = \omega t$ and $F_y = m_e\omega^2\epsilon \sin \omega t$.

If the structure supporting the machine which generates the unbalanced rotating force is sufficiently flexible in the y direction, it will respond appreciably to the periodic force $m_e\omega^2\epsilon \sin \omega t$. If, in addition, it is relatively rigid in the x direction and hardly responds at all to the component F_x, it can be modelled as a single degree-of-freedom system, for which we can write

$$m\ddot{y} + c\dot{y} + ky = m_e\omega^2\epsilon \sin \omega t.$$

This is obviously of the same form as equation (4) with F set equal to $m_e\omega^2\epsilon$, a magnitude independent of time t. By comparison with equation (7), the particular integral part of the solution is

$$y = \frac{m_e\omega^2\epsilon}{\sqrt{[(k-m\omega^2)^2+(c\omega)^2]}} \sin(\omega t + \phi)$$

where   (9)

$$\tan \phi = \frac{-c\omega}{k-m\omega^2}.$$

In Section 3.3 we were interested in how the amplitude of the vibration might alter, and we plotted it against excitation frequency (Figure 6(a)), presuming this to be a variable parameter for a given system. However, if we do this here we shall not get the same result because the amplitude of the exciting force is now no longer independent of excitation frequency: it is proportional to the square of it. Indeed if we just consider the situation where the rotational (excitation) frequency is very low, the centripetal force will be extremely small and so the amplitude of vibration will be very small. On the other hand, as the rotational frequency reaches very high values, $m\omega^2$ will be much larger than k and also much larger than $c\omega$. Therefore the denominator in equation (9)

$$\sqrt{[(k-m\omega^2)^2+(c\omega)^2]}$$

will be approximately $m\omega^2$ and so y will approach

$$\frac{m_e\omega^2\epsilon}{m\omega^2} = \frac{m_e\epsilon}{m}.$$

That is the amplitude of vibration becomes asymptotic to a non-zero value.

The full variation of amplitude with frequency for several values of damping ratio is shown in non-dimensionalized form in Figure 9 and should be compared with the curves of Figure 6(a). The non-dimensionalized expression is

$$\frac{Ym}{\epsilon m_e} = \frac{(\omega/\omega_n)^2}{\sqrt{\left\{\left[1 - \left(\frac{\omega}{\omega_n}\right)^2\right]^2 + \left(\frac{2\alpha}{\omega_n} \times \frac{\omega}{\omega_n}\right)^2\right\}}} .$$

The phase diagram corresponding to Figure 9 is identical with Figure 6(b).

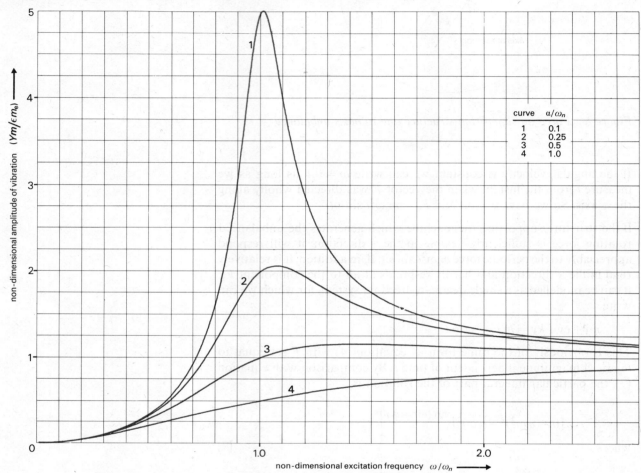

curve	α/ω_n
1	0.1
2	0.25
3	0.5
4	1.0

Figure 9   Amplitude as a function of excitation frequency when excitation is due to mass unbalance

SAQ 4

A machine is supported by a spring mounting with viscous damping. The static deflection, due to the weight of the machine, is 2.0 mm.

(a) Calculate the undamped natural angular frequency.

(b) When the machine is working, an unbalanced rotating mass produces a vertical alternating force at a frequency equal to the speed of the driving shaft. It is found that the steady-state amplitude of forced vibration is the same at a driving shaft speed of 61.3 rad s^{-1} as it is at a driving shaft speed of 107.6 rad s^{-1}. Use the graph of Figure 9 to estimate the damping ratio of the machine on its mounting.

(c) If the steady state amplitude at these two speeds is 0.9 mm and the rotating mass is one twentieth of the total mass of the machine, use Figure 9 to estimate the eccentricity ϵ of the rotating mass.

(d) What will be the phase angle between the vertical alternating force and the steady state motion of the machine at 107.6 rad s^{-1}? (Use Figure 6(b)).

3.5 Transient vibration

The complementary function part of the solution of the forced vibration differential equation (the exponential term of the right-hand side of equation (8)) is the single degree-of-freedom representation of transient vibration. Although it is convenient to be able to separate mathematically the transient and steady-state components, this is in many ways an artificial distinction which tends to give a false impression of the variation with time of the vibration. It arises because, in order to understand the nature of the complementary function, it is conventional to plot the amplitude as a function of time t, but for understanding the nature of the particular integral the relevant graph is of amplitude versus excitation frequency, this amplitude being independent of time.

To combine the effect of these two parts of the total solution, it is interesting to plot the total displacement (sum of particular integral and complementary function) against time for several values of the parameters. Basically two different types of behaviour can be identified, one caused when the excitation frequency is greater than the system natural frequency and one when it is less than the natural frequency. Typical curves are shown in Figures 10(a) and 10(b) respectively.

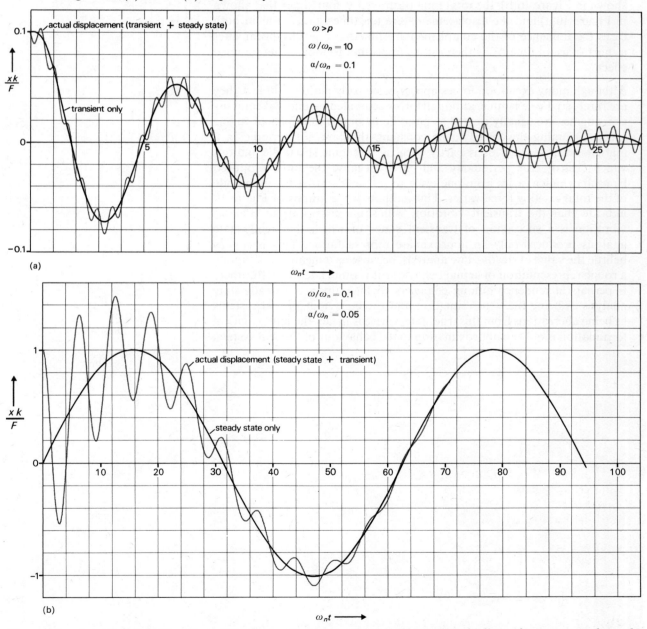

Figure 10   *Amplitude as a function of excitation frequency showing the combined effect of transient and steady-state components (a)* $\omega > p$ *(b)* $\omega < p$

27

These curves were obtained by suddenly applying a sinusoidal force of circular frequency ω to the system and recording the subsequent displacement of the mass. Separate plots of the transient component (in Figure 10(a)) and of the steady-state component (in Figure 10(b)) have been added in order to illustrate the fact that the actual motion eventually becomes indistinguishable from the steady-state component. The shape of the earlier parts of each of these graphs depends on the exact initial velocity and displacement as well as on the parameters of the vibrating system and of the exciting force. (Note that transient vibrations of the system will be excited whenever there is a sudden change in the exciting force.)

When ω is greater than p (Figure (10(a)), the mean of the higher frequency vibration itself fluctuates in a decaying manner. The black curve representing this fluctuation of the mean is the complementary function. When ω is less than p (Figure 10(b)), it is the vibration whose mean fluctuates which decays, ultimately leaving only the mean level (the black line) which in this case is the particular integral. In addition to the obvious differences of shape between the curves there is also a difference in coordinate values. For the case of a low frequency of the exciting force shown in Figure 10(b), the total time portrayed is four times that shown in Figure 10(a) and the displacements are ten times larger. But in both cases the influence of the complementary function does disappear after a short space of time. Why then is it so important to look at the summed effect?

Although many excitation forces appear to be truly periodic, in fact they are rarely so. For example, although forces generated by internal combustion engines are usually treated as periodic in simple multiples of engine speed, assuming this speed to be constant, they are not periodic in a mathematical sense. This is partly because of engine speed fluctuations which destroy true periodicity and partly because the force produced by each cylinder varies from one cycle to the next so that the amplitude of the total exciting force varies. Thus, although the graphs of Figure 10 indicate that the transient vibration will soon disappear, in practice vibration is a succession of transient states of which the steady-state analysis produces only an approximate representation. This is not to belittle the value of steady-state analysis, because as a means of obtaining a rough representation of actual behaviour it is quite adequate. But there is perhaps a tendency among engineers to look upon the steady-state analysis as being the 'true' situation and the behaviour of real systems as being aberrations from this true state, whereas in fact the mathematical formulation is only an approximation to the behaviour of the real system.

TWO DEGREE-OF-FREEDOM SYSTEMS

Many systems and devices are too complex to be adequately represented and modelled by a single degree-of-freedom system. Our main evidence for this is that many exhibit more than one natural frequency. This means that there is more than one value of excitation frequency near which the amplitude of vibration will build up to a maximum and then fall away as the frequency is passed.

The easiest way in which we can begin to investigate the problem is by 'doubling-up' our schematic diagram for a single degree-of-freedom system (Figure 11). Let the system be excited into vibration by the force $F_1 \sin \omega t$ acting on the body of mass m_1 as shown. We can then write an equation of motion for each body.

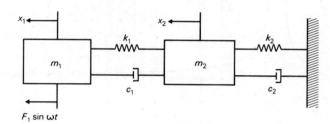

Figure 11   Representation of a two degree-of-freedom system

The mass m_1 is acted on by the external force and by the forces in the spring k_1 and damper c_1. The force in the spring is proportional to the extension of the spring which is $x_1 - x_2$. If $x_1 < x_2$ the extension is negative (compression), and so the spring force is negative (compressive). Thus the spring force is $k_1(x_1 - x_2)$ and in a similar way the damper force is $c_1(\dot{x}_1 - \dot{x}_2)$. Thus the equation of motion for the mass m_1 is

$$m_1 \ddot{x}_1 = F_1 \sin \omega t - k_1(x_1 - x_2) - c_1(\dot{x}_1 - \dot{x}_2).$$

The body of mass m_2 is affected by the forces in both springs and damper. That is

$$m_2 \ddot{x}_2 = k_1(x_1 - x_2) + c_1(\dot{x}_1 - \dot{x}_2) - k_2 x_2 - c_2 \dot{x}_2.$$

Rearranging these equations gives

$$(m_1 \ddot{x}_1 + c_1 \dot{x}_1 + k_1 x_1) - (c_1 \dot{x}_2 + k_1 x_2) = F_1 \sin \omega t$$

and
$$-(c_1 \dot{x}_1 + k_1 x_1) + m_2 \ddot{x}_2 + (c_1 + c_2)\dot{x}_2 + (k_1 + k_2)x_2 = 0. \tag{10}$$

In this course we do not intend to investigate the solution of these equations in depth. It is sufficient to say that they show that when the damping is light there are two resonant frequencies, and that these do not coincide with the two undamped natural frequencies of the separate parts $\sqrt{(k_1/m_1)}$ and $\sqrt{(k_2/m_2)}$. The way in which the displacement amplitudes X_1 and X_2 of the steady-state responses of the two bodies vary as a function of excitation frequency is shown in Figure 12 for zero damping $(c_1 = c_2 = 0)$. Two points are worth noting:

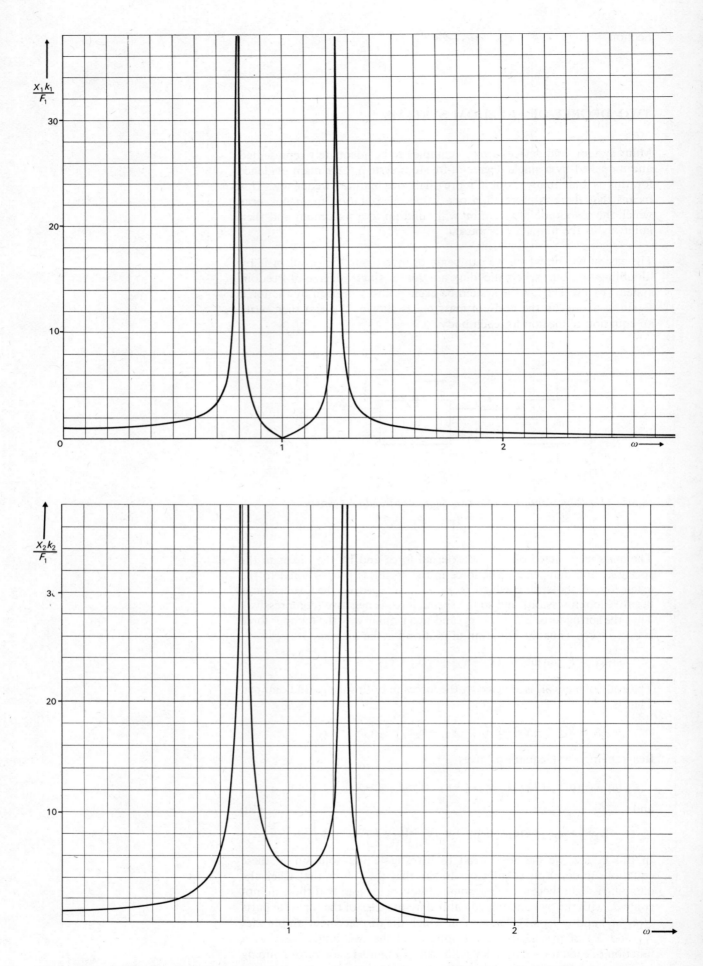

Figure 12   Amplitudes of displacement as a function of excitation frequency for a two degree-of-freedom system

(a) that the system has two resonant frequencies, near each of which the displacements of both masses are large;

(b) that there is a frequency between the two resonant frequencies at which the amplitude of the mass m_1 is zero, that is m_1 is stationary.

The equations (10), which are coupled together (dependent) through the terms x_1, $\dot{x}_1$, x_2 and $\dot{x}_2$, are appreciably more complex than that for a single degree-of-freedom system (equation (4)) and, provided damping is small, they represent a system with only two resonant frequencies. Many real systems exhibit ten or more resonant frequencies within the range of excitation frequencies of interest and if we approach these in a similar manner we obtain ten (or more) *simultaneous* differential equations. This is more than enough to make us pause in our hitherto straightforward use of mathematics and mathematical models, and look for other ways of finding out about the vibration characteristics of systems.

Section 5   *Not Examinable.*

SUMMARY

1   To look at the effects of time-dependent exciting forces on the vibration of a single degree-of-freedom system, we concentrate on the effect of a single sinusoidally varying force. This is because nearly all periodic forces can be expressed as a sum of sine wave components.

2   The equation to be solved is therefore

$$m\ddot{x} + c\dot{x} + kx = F\sin\omega t$$

and it has a complete solution

$$x = e^{-\alpha t}(A\sin pt + B\cos pt) + \frac{F/k}{\sqrt{\left[\left(1 - \frac{\omega^2}{\omega_n^2}\right)^2 + \left(\frac{2\alpha}{\omega_n} \times \frac{\omega}{\omega_n}\right)^2\right]}}\sin(\omega t + \phi)$$

where A and B are dependent on initial conditions and

$$\phi = \tan^{-1}\left(\frac{\frac{-2\alpha}{\omega_n} \times \frac{\omega}{\omega_n}}{1 - \frac{\omega^2}{\omega_n^2}}\right).$$

3   It is convenient to think of the system properties in terms of the dimensionless parameters ω/ω_n and Xk/F, and the damping ratio α/ω_n, where $\alpha = c/2m$ and $\omega_n = \sqrt{(k/m)}$. Note also that the damping ratio α/ω_n is equal to c/c_c where c_c is the *critical* damping coefficient.

4   The total solution is plotted for specific cases in Figure 10. The amplitude of the steady-state vibration only is shown in Figure 6(a) for the case where F is independent of ω and in Figure 9 for the case when F is proportional to ω^2. The phase relationship for both cases is shown in Figure 6(b).

Part II

Dimensional Analysis

INTRODUCTION

At the end of Part I we looked briefly at the equations of vibration for a two degree-of-freedom system and the meaning of their solution. If we try to analyse the behaviour of more complex real structures such as heavy beams, plates or cylinders, the mathematics becomes more difficult and might well be beyond our capability. In such cases there is a tendency to turn to an empirical approach in which we would hope to measure all the characteristics of interest.

If we are only concerned with one particular beam, say, and shall never again want similar information, then the beam can be tested, its response measured and the job is finished. However, if we are trying to determine some *general* behaviour characteristics, for example how the frequency of a certain mode of vibration of a cylinder depends on the length and diameter, we must conduct a number of experiments which, for economy, should be planned to give the required information with the fewest possible tests. The technique of dimensional analysis enables us to do this.

Dimensional analysis may also be used in a different situation. A particular structure might be so large and expensive (a suspension bridge, a cooling tower or an aircraft fuselage) that it is not feasible to build a prototype just for vibration testing. The testing of a model is an obvious, less costly alternative. However, it is a common misconception that a model for dynamic testing is nothing more than a well-executed geometric scale model. This is far from the truth: many toys which are scaled replicas of full-size machinery do not *behave* like their full-size counterparts. Model trains and their tracks are not wrecked every time a derailment occurs; balsa-wood structural models can be dropped to the floor and not be damaged; a film company's model simulating the bursting of a huge dam wall just does not look right.

Measurements on a model can only be related to the corresponding quantities for its full-scale prototype when the two systems are *dynamically similar*, and it is in obtaining this similarity that dimensional analysis is particularly useful.

The whole idea of similarity is discussed in detail in Section 7 of this part of the unit. Section 8 goes on to look at the application of dimensional analysis in investigating the behaviour of some particular systems. Section 9 illustrates how it might be used in making a useful scale model of a complex structure.

PHYSICAL SIMILARITY

Physical similarity is based on the quantitative correspondence of two or more physical systems. When systems are similar and are behaving similarly, the quantitative relations representing the properties of one can be derived from those holding for another. As we have seen, this is of particular value in relating measurements on a model with the corresponding quantities for its full-scale prototype.

7.1 Basic dimensions and geometric similarity

In Unit 1 it was shown that all quantities have dimensions as well as magnitude. Just as the magnitude can be derived from various fundamental relations in which the quantity appears, so also can the dimensions. All dimensions can be expressed ultimately in terms of a few special ones which are referred to as basic dimensions. In mechanics we usually consider just three dimensions as basic, although in thermodynamics some others may have to be introduced.

The basic mechanical dimensions are the dimension of length, the dimension of mass and the dimension of time. These are signified by $[L]$, $[M]$ and $[T]$ respectively. A moment of thought will show that kinematics, being concerned only with motion and not with the forces required to produce or to change motion, is related only to $[L]$ and $[T]$ (in Units 2 and 3 of this course we had no need to introduce the concept of mass at all). Dynamics is concerned with all three dimensions. A geometric scale model, however, is concerned only with $[L]$, so such a model cannot, except by accident, give a scaled reproduction or representation of kinematic or dynamic processes.

The important factor about such a *geometrically similar* model is that its size can be specified by a single parameter. For example, in a kit for constructing a model aircraft which is 1/72 of full scale, each dimension of the full scale aircraft is reduced by a factor of 1/72 in the model. Another way of describing the relationship between model and full scale aircraft would be to specify the actual size of some *characteristic length* of the model, say the wing-span, and then all the dimensions would be scaled down by the same ratio: that of wing-span of model to wing-span of full-scale aircraft. (The full-scale device is often denoted by the term *prototype*. Conversationally, this word is usually used when the original device being modelled is a single new assembly or component. However, here we use the word in the more general sense to distinguish a device from its model.)

geometric similarity

Thus when geometric similarity exists between two systems, the scale factor (ratio of corresponding sizes) or the magnitude of a characteristic length of one system is sufficient to define the size relationships.

Exercise

A solid disc is 30 mm thick and has a radius of 450 mm. Its mass m_1 is 15.0 kg and it is able to rotate about an axis through its centre as shown in Figure 13. Its second moment of mass about this axis is equal to

$$\tfrac{1}{2}m_1 r_1{}^2 = \tfrac{1}{2} \times 15 \times (0.45)^2 = 1.52 \, \text{kg m}^2.$$

What would be the corresponding value of the second moment of mass of a geometrically similar disc of half the size? Assume that the discs are made of the same material.

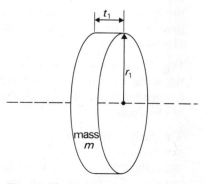

Figure 13

For a disc made of material of density ρ,

$$m = \rho \pi r^2 t$$

and

$$I = \tfrac{1}{2} m r^2$$

about the required axis, therefore

$$I = \rho \frac{\pi}{2} r^4 t.$$

Thus, using the subscript 1 to refer to the full-size disc and 2 for the half-size disc, we have

$$I_1 = \rho \frac{\pi}{2} r_1{}^4 t_1,$$

$$I_2 = \rho \frac{\pi}{2} r_2{}^4 t_2$$

thus

$$\frac{I_2}{I_1} = \left(\frac{r_2}{r_1}\right)^4 \left(\frac{t_2}{t_1}\right).$$

But $\quad r_2 = \tfrac{1}{2} r_1 \quad$ that is $\quad r_2/r_1 = \tfrac{1}{2}$
and $\quad t_2 = \tfrac{1}{2} t_1 \quad$ that is $\quad t_2/t_1 = \tfrac{1}{2}.$

Therefore

$$\frac{I_2}{I_1} = \frac{1}{2^5} = \frac{1}{32},$$

$$I_2 = \frac{I_1}{32} = 0.0475 \, \text{kg} \, \text{m}^2.$$

7.2 Kinematic similarity

Kinematic similarity concerns systems in motion. For two systems to be kinematically similar they must first be geometrically similar and second the paths traced out during motion by geometrically similarly placed points in the two systems must also be geometrically similar. Not an easy definition to follow, so what does it mean?

Imagine that a motor car travelling round a bend skids and turns round. If we want to make a model car behave in the same way we must first ensure that the model car and the bend in the model road are geometrically similar to full-size versions. Having done that, the skidding of the model car is kinematically similar to that of the full-size car only when the paths traversed by every point of the skidding model are geometrically similar to the paths traversed by corresponding points on the full-size car. If this is the case, then the speeds of all points of the model at any instant are proportional to the speeds of corresponding points on the full-size car and the directions of the movement are identical in the two cases.

Figure 14(a) shows the paths followed by a shot fired from a cannon when the cannon is set to different angles of elevation. Figure 14(b) shows a much smaller cannon with the path taken by a shot for just one angle of elevation.

The paths 3 and 4 are geometrically similar but the paths 1, 2 and 4 are geometrically dissimilar. The muzzle velocity in Figure 14(a) is three times the muzzle velocity in Figure 14(b) and so it follows (since 3 and 4 represent kinematically similar conditions) that the shot velocity at the

36

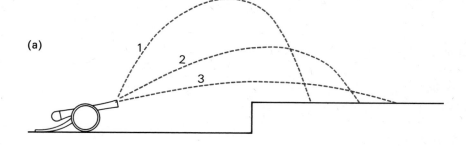

(a)

(b)

Figure 14 Paths taken by shots from two geometrically similar cannons; the paths 3 and 4 represent kinematically similar conditions

highest point of the curve 3 in Figure 14(a) is three times the velocity at the highest point in curve 4, that the shot hits the ground with three times the velocity, that it travels three times as far before hitting the ground but that it strikes the ground at the *same* angle. (The step is assumed to be three times as high.)

It is important to recognize that the cannon in Figure 14(a) is only about twice as large as that in Figure 14(b). This is an example of the fact that the ratio for geometric similarity of the physical system (close to 2 : 1 in this case) need not be the same as the ratio for kinematic similarity which is 3 : 1. They are two separate factors, one associated with the physical size of the objects and the other associated with motion, that is the rate at which displacements take place. Therefore, geometric similarity involves only the dimension $[L]$, or the dimensionless ratio l_M/l_P, where l_M is a characteristic length in a model and l_P is the corresponding length in the prototype. But kinematic similarity involves the dimensions $[L]$ and $[T]$, or the dimensionless ratio

$$\frac{l_M}{t_M} \bigg/ \frac{l_P}{t_P} \quad \text{that is} \quad \frac{l_M t_P}{l_P t_M} \quad \text{or} \quad \frac{l_M}{l_P} \bigg/ \frac{t_M}{t_P}.$$

Hence, kinematic similarity cannot exist unless geometric similarity also exists.

Exercise

The mechanism of a reciprocating internal combustion engine is shown in Figure 15. The crank OA rotates continuously about O with a constant angular velocity of 2400 rev min^{-1} while the piston B reciprocates. The crank has a length of 80 mm and the connecting-rod AB is 200 mm long. The piston has a length of 76 mm and a diameter of 60 mm. The slider is in line with the crank.

(a) What dimensions must a similar mechanism have if it is to be geometrically similar and if the crank is to be 20 mm long?

(b) What conditions must be satisfied for kinematic similarity of the two mechanisms?

(c) If the angular velocity of the connecting-rod for the smaller mechanism is 200 rad s^{-1} at its dead-centre positions, what is the ratio of velocities between the two mechanisms?

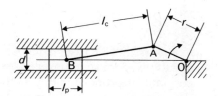

Figure 15

(a) Denote the given mechanism by the subscript 1 and the second smaller mechanism by the subscript 2. Given that $r_2 = 20$ mm and $r_1 = 80$ mm,

$$r_1/r_2 = 4.$$

Thus all dimensions in the second mechanism are $\frac{1}{4}$ of the corresponding dimensions in the larger mechanism:

$$l_{c2} = \tfrac{1}{4} \times 200 = 50 \text{ mm}$$
$$l_{p2} = \tfrac{1}{4} \times 76 \;= 19 \text{ mm}$$
$$d_2 = \tfrac{1}{4} \times 60 \;= 15 \text{ mm}.$$

(b) Since the motion of a constrained mechanism is pre-determined by its geometry, satisfaction of geometrical similarity in these cases automatically gives kinematic similarity, provided that the crank of the second mechanism also rotates at constant angular velocity.

(c) At the dead-centre position, the absolute velocity of point B (the velocity of the piston) is zero and so the velocity of the pin-joint A is $\omega_c l_c$ where ω_c is the con-rod angular velocity. But the velocity of joint A is also $\omega_r r$ where ω_r is the crank angular velocity. Therefore as $\omega_{c2} = 200 \text{ rad s}^{-1}$,

$$\omega_{r2} = \omega_{c2} \times \frac{50}{20} = 500 \text{ rad s}^{-1}.$$

Thus the ratio of angular velocities is

$$\frac{\omega_{r1}}{\omega_{r2}} = 2400 \times \frac{2\pi/60}{500} = 0.503 \simeq \frac{1}{2}$$

and the ratio of the velocities of every two corresponding points in the original slider–crank and the model is equal to this. For example,

$$\frac{v_{A2}}{v_{A1}} = \frac{\omega_{c2} l_{c2}}{\omega_{r1} r_1} = \frac{\omega_{c2} l_{c2}}{2400 \times (2\pi/60) \times 80} = \frac{200 \times 50 \times 60}{2400 \times 2\pi \times 80} = \frac{25}{16\pi} \simeq \frac{1}{2}.$$

The two ratios we have just calculated are so nearly equal to $\frac{1}{2}$ that we will treat them as equal to $\frac{1}{2}$ when considering this exercise further.

We have shown that kinematic similarity is related to both similarity of velocities and to geometric similarity. But acceleration is also a kinematic parameter and we need to know what relationships hold for this.

Consider the engine mechanism of the preceding exercise. The centripetal acceleration of the pin A is given by $\omega_r^2 r$. That is the ratio of accelerations between the two mechanisms is given by

$$\left(\frac{\omega_{r1}}{\omega_{r2}}\right)^2 \left(\frac{r_1}{r_2}\right) = 1$$

in this particular case. Thus we can see that in kinematically similar situations, corresponding accelerations are proportional, and that the constant of proportionality is the product of the size scale factor and the square of the angular velocity scale factor.

The acceleration ratio could have been expressed using a linear velocity scale factor, say v_1/v_2, giving

$$\frac{a_1}{a_2} = \frac{v_1^2}{r_1} \bigg/ \frac{v_2^2}{r_2} = \left(\frac{v_1}{v_2}\right)^2 \left(\frac{r_2}{r_1}\right).$$

The basic dimensions for this are given by

$$\frac{a_1}{a_2} = \left(\frac{v_1}{v_2}\right)^2 \left(\frac{r_2}{r_1}\right) = \left[\frac{L_1 T_1^{-1}}{L_2 T_2^{-1}}\right]^2 \left[\frac{L_2}{L_1}\right] = \left[\frac{L_1}{L_2}\right]\left[\frac{T_2}{T_1}\right]^2.$$

In the slider–crank example we found that $a_1/a_2 = 1$. So what time dimension ratio $[T_1/T_2]$ is required to maintain kinematic similarity?

We can write

$$\left[\frac{L_1}{L_2}\right] \bigg/ \left[\frac{T_1}{T_2}\right]^2 = 1$$

therefore

$$\left[\frac{T_1}{T_2}\right]^2 = \left[\frac{L_1}{L_2}\right] = 4,$$

giving

$$\left[\frac{T_1}{T_2}\right] = 2.$$

To answer the same question, we could have reverted to the ratio

$$\frac{\omega_{r1}}{\omega_{r2}} = \frac{1}{2}$$

in which, as the dimensions of ω are only $[T^{-1}]$, we have a direct statement that

$$\left[\frac{T_1^{-1}}{T_2^{-1}}\right] = \frac{1}{2},$$

therefore

$$\left[\frac{T_2}{T_1}\right] = \frac{1}{2} \quad \text{or} \quad \left[\frac{T_1}{T_2}\right] = 2.$$

Alternatively, consider the velocity ratio

$$\frac{v_{A1}}{v_{A2}} = 2.$$

This gives

$$\left[\frac{L_1 T_1^{-1}}{L_2 T_2^{-1}}\right] = 2,$$

that is

$$\left[\frac{L_1}{L_2}\right] \bigg/ \left[\frac{T_1}{T_2}\right] = 2,$$

$$\left[\frac{T_1}{T_2}\right] = \frac{1}{2} \times \left[\frac{L_1}{L_2}\right] = 2.$$

Hence a time interval of 1 s in the original mechanism corresponds to a time interval of 0.5 s in the (geometrically and) kinematically similar model.

7.3 Dynamic similarity

If the ratio of the forces on two kinematically similar systems is the same for all corresponding pairs of forces, these systems are said to be *dynamically similar*. Since $F = ma$, the external forces (dimensions $[MLT^{-2}]$) on a system must be scaled so that their ratio is the product of the mass (dimensions $[M]$) and acceleration (dimensions $[LT^{-2}]$) ratios. The new factor here is the mass.

dynamic similarity

Thus in looking at similarity, we have three fundamental dimensions: length $[L]$, time $[T]$, and mass $[M]$. Where we have proportionality of length between systems, the systems are said to be geometrically similar. If, in addition, we have proportionality of time (inferred by a constant velocity ratio) the systems are kinematically similar. If we also have proportionality of mass (and hence of force), we have the ultimate in similarity, dynamic similarity. Dynamic similarity is sometimes expressed as *complete similarity* because it requires both kinematic and geometric similarity as pre-conditions. The remainder of this unit is about complete similarity.

complete similarity

Exercise

Consider a prototype machine having a characteristic length l_P, a characteristic mass m_P and velocity v_P. This is to be studied using a model which is dynamically similar to it. The model is made of the same material but is one-quarter of the full-scale size and it has velocities which are one-quarter of the corresponding velocities in the prototype. What will be the relationship between forces measured in the model and those existing in the prototype?

If a prototype length is l_P, the corresponding model length is $l_M = l_P/4$. That is, the length ratio

$$l_R = l_P/l_M = 4.$$

If the density of the material of which the prototype is made is ρ_P, then $\rho_P = \rho_M$ and

$$\rho_R = \rho_P/\rho_M = 1.$$

A component of the prototype has a mass $m_P \propto \rho_P l_P^3$. Hence

$$m_R = m_P/m_M = \rho_R l_R^3 = 1 \times 64 = 64.$$

If a characteristic velocity of a component in the prototype is v_P, a characteristic time interval is

$$t_P = l_P/v_P$$

and in the model

$$t_M = l_M/v_M.$$

Therefore

$$t_R = \frac{l_P}{l_M} \times \frac{v_M}{v_P} = \frac{l_R}{v_R}.$$

But $v_R = v_P/v_M = 4$ and $l_R = l_P/l_M = 4$, hence

$$t_R = 1.$$

A characteristic acceleration in the prototype

$$a_P = \frac{l_P}{t_P^2},$$

and it follows that in the model,

$$a_M = \frac{l_M}{t_M^2}.$$

Hence

$$a_R = \frac{a_P}{a_M} = l_R t_R^{-2} = 4.$$

A characteristic force in the prototype F_P is given by

$$F_P = m_P a_P$$

and in the model,

$$F_M = m_M a_M.$$

Therefore

$$F_R = m_R a_R = 64 \times 4 = 256.$$

That is, a force of 1 N measured in this dynamically similar model corresponds to a force of 256 N in the prototype.

SAQ 5

SAQ 5

The stopping distance of a vehicle is to be investigated by tests on a similar model which is one-tenth full size. The material of the model is the same as that of the prototype. The (constant) braking force on the model is one-fiftieth of that on the prototype. Calculate the ratio of the initial velocity of the model to that of the prototype, the ratio of corresponding time intervals, and the ratio of stopping distances.

THE TECHNIQUE OF DIMENSIONAL ANALYSIS

In mechanical engineering, we often have to investigate the way in which the performance of a complex system varies as different parameters of it are changed. As already indicated, the technique of dimensional analysis can frequently point the way to the necessary form of experimental work. It allows the variation in several parameters to be accommodated on single graphs, using non-dimensional parameters or groups of variables as ordinates and abscissae. For theoretically insoluble problems such experiments are necessary to find the numerical values of the constants involved.

One example is the aerodynamic excitation of simultaneous bending and twisting of a structure. This is quite a common problem, particularly in considering the design and construction of long span bridges. The well known collapse of the Tacoma Narrows Bridge in the USA in 1940 can be attributed to this. So also can the collapse of the cooling towers at Ferrybridge in England in 1965.

8.1 A cantilever beam

We shall use dimensional analysis to investigate the cantilever beam that you met in Units 4 and 5. You should be familiar with the bending moments, shear forces and resulting stresses that occur at each section of a beam, and you should be able to calculate them or express them in terms of the loads applied to the beam and its dimensions. In Unit 5 we also discussed a theoretical means by which deflections of simple beams may be calculated. Let us now apply the technique of dimensional analysis to see what we can find out about the deflection of a beam loaded in a specific way.

Consider the cantilever in Figure 16. We want to find the form of the expression for the deflection of the cantilever under the point load.

The first question to ask is which physical quantities determine the deflection.

By intuition, the two parameters of load W and span l must be involved. In addition, from the work in Unit 5, we have good reason for selecting the Young's modulus E for the material of the beam as well as the second moment of area I_A of the cross-section of the beam. Young's modulus is of consequence because a steel beam would deflect far less than a rubber one (steel has a relatively high value of E and rubber a low value). The second moment of area is important because a thick beam would deflect less than a thin beam of the same material.

These four quantities are all particularly relevant to the problem, and so we can make the general statement that the deflection is a function of W, l, E and I_A. That is,

$$y = f(W, l, E, I_A). \tag{11}$$

(Note that f is not an algebraic constant or variable denoting a quantity with dimensions. It is merely an *operator* – an indicator of some operation that could be performed upon the terms inside the bracket. The symbol y has been used for deflection, rather than the symbol v which was used in Unit 5, to prevent confusion with the velocity variable.)

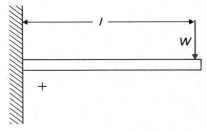

Figure 16   Loaded cantilever

We now want to find the correct arrangement of the variables on the right-hand side of equation (11).

Let us begin by expressing the variables on both sides in terms of the basic dimensions $[M]$, $[L]$ and $[T]$:

$$[L] = \phi[MLT^{-2}, L, ML^{-1}T^{-2}, L^4].$$

Here

$$\phi[MLT^{-2}, L, ML^{-1}T^{-2}, L^4] \equiv [MLT^{-2}]^a \times [L]^b \times [ML^{-1}T^{-2}]^c \times [L^4]^d$$

that is, it denotes the product of powers of the groups of dimensions inside the square brackets.

We now make the dimensions on each side of equation (11) tally: we arrange the four groups of dimensions on the right-hand side so that their combined dimension is $[L]$. By trial and error we find that we can write

$$[L] = \left[\frac{(MLT^{-2}) \times (L)^3}{(ML^{-1}T^{-2}) \times (L^4)} \right],$$

(that is $a = 1$, $b = 3$ and $c = d = -1$). Thus we get the expression

$$y = f_1\left(\frac{Wl^3}{EI_A} \right).$$

This expression is *dimensionally correct* or *dimensionally homogeneous* (that is, the dimensions are the same on both sides of the equation) and it is the nearest approach to a solution that dimensional analysis will provide. Dimensional analysis does not enable us to determine what f_1 is. This must be done by experiment or further analysis.

In fact when the appropriate expression for the deflection of the end of the cantilever is derived by a mathematical analysis, the complete expression for the deflection turns out to be

$$y = \frac{Wl^3}{3EI_A}.$$

8.2 A vibrating beam

Let us now take a simple dynamic example. What can we find out about the natural frequencies of vibration of a uniform beam of rectangular section which is simply supported at both ends?

First, the frequencies will depend on the length of the beam (l) and on the size of its cross-section ($b \times d$). Second, the vibration is an oscillating exchange of energy between the forms of strain and kinetic energy. At one instant the system has kinetic energy by virtue of its moving mass and at another instant it has strain energy by virtue of its stiffness. Thus the frequencies of vibration are also dependent on these parameters ($\omega_n = \sqrt{(k/m)}$, for example).

Let the mass be characterized by the density ρ and the stiffness by Young's modulus E. Then we can write

$$\omega = f(l, b, d, \rho, E) \tag{12}$$

and, looking at the dimensions of this equation,

$$\left[\frac{1}{T} \right] = \phi\left[L, L, L, \frac{M}{L^3}, \frac{M}{LT^2} \right]. \tag{13}$$

Exercise

Arrange the groups of dimensions on the right-hand side of equation (13) so that this side also has dimensions of $[1/T]$. Hence write equation (12), $\omega = f(l, b, d, \rho, E)$ in dimensionally correct form.

The only variable containing the dimension $[T]$ is the modulus of elasticity E. Therefore E will have to be contained in the set of groups that we form on the right-hand side. In fact we can see that E will have to be in the form of its square root in order to achieve a dimension of $[1/T]$:

$$\sqrt{E} = \left[\left(\frac{M}{L}\right)^{1/2} \times \frac{1}{T}\right].$$

However, this introduces the dimensions $[M]$ and $[L]$ in the form $[\sqrt{(M/L)}]$, and the only term we can combine with $\sqrt{E}$ in order to cancel out the unwanted $[\sqrt{M}]$ is $1/\sqrt{\rho}$:

$$\frac{1}{\sqrt{\rho}} = \left[\frac{L^{3/2}}{M^{1/2}}\right].$$

Now we have

$$\sqrt{\frac{E}{\rho}} = \left[\frac{M^{1/2}}{L^{1/2}} \times \frac{1}{T} \times \frac{L^{3/2}}{M^{1/2}}\right] = \left[\frac{L}{T}\right].$$

Therefore multiplying the term $\sqrt{(E/\rho)}$ by $1/l$ (or by $1/b$ or $1/d$) will give us a set of groups with dimensions $[1/T]$. We can now write the equation for frequency in the dimensionally correct form:

$$\omega = f_1\left(\frac{1}{l}\sqrt{\frac{E}{\rho}}, \frac{1}{b}\sqrt{\frac{E}{\rho}}, \frac{1}{d}\sqrt{\frac{E}{\rho}}\right).$$

In this exercise we derived one dimensionally correct form of equation (12). Notice however that all the groups of dimensions in the brackets can be made up of the following groups:

$$\frac{1}{l}\sqrt{\frac{E}{\rho}}, \quad \frac{b}{l}, \quad \frac{d}{l}.$$

For example

$$\frac{1}{b}\sqrt{\frac{E}{\rho}} = \left(\frac{1}{l}\sqrt{\frac{E}{\rho}}\right) \times \left(\frac{1}{b/l}\right).$$

Thus another way to write our equation for ω is to use these three groups. Notice, however, that we cannot put them all into the f_1 bracket because they have different dimensions: b/l and d/l are dimensionless and $(1/l)\sqrt{(E/\rho)}$ has dimensions of $[1/T]$. But we can get over this by writing

$$\omega l \sqrt{\frac{\rho}{E}} = f_1\left(\frac{b}{l}, \frac{d}{l}\right). \tag{14}$$

Both sides of equation (14) are dimensionless, and this is a useful form for the equation as it is not dependent on any particular system of units. We have now got as far towards a solution of the problem as dimensional analysis will take us.

The complete expression can be found by conducting a series of experiments in which the frequencies of vibration of two ranges of rectangular section, simply supported beams are measured, one range having b/l constant and the other having d/l constant. It shows that the frequencies are, in fact, independent of b/l and are linearly dependent on d/l. That is,

$$\omega l \sqrt{\frac{\rho}{E}} = \text{constant} \times \frac{d}{l}.$$

A heavy beam has an infinite number of natural frequencies and they are all given by an equation of this form, but all of them have a different characteristic value of the constant.

The result of our analysis, equation (14), is not restricted to simply supported beams, since although we had this type of support in mind, it did not feature in our list of variables. The list would have been just the same if we had been thinking of rectangular section cantilever beams or of rectangular beams with any other type of support. For example, a beam which is rigidly built in at both ends has a lowest natural frequency which is given by

$$\omega l \sqrt{\frac{\rho}{E}} = 2.056\pi \frac{d}{l}.$$

An important point to notice about the dimensional analysis we have just done is that although we started with six variables (including ω), we ended up by expressing ω in terms of only three dimensionless groups. To make our procedure of analysis more methodical in the examples which follow, we shall make use of a rule which tells us the required number of dimensionless groups in terms of the number of variables originally specified. This rule is discussed in more detail in Appendix B (and you should read this before starting Section 8.3 if you have time). We can state it with sufficient accuracy for our present purpose by saying that the number of dimensionless groups we use is equal to the original number of variables (including the one on the left-hand side of the equation) less the number of fundamental dimensions ($[M],[L],[T]$) needed to build up the dimensions of all the variables.

Armed with this rule, we can now go on to a more complex problem.

8.3 Vibrations of a structure

Let us return to the type of structure referred to at the beginning of this section where the vibration is aerodynamically excited. The frequencies of vibration of such a structure can be considered to be a function of the size of the structure (l), the moduli of elasticity and rigidity (E and G) of the material (governing the bending and shearing respectively), and the density (ρ_m) of the material governing the structure's mass. The frequencies are also a function of the parameters of the incident air stream: the air velocity v, air density ρ_a and the viscosity μ. The problem is to find the *form* of the relationship between the frequency and the other variables.

(You need not worry about a formal definition of viscosity here. It is sufficient at this stage to understand that all fluids have the property of being viscous and that the amount of viscosity exhibited by different fluids varies over a wide range. There are many familiar examples of this: different grades of motor oil have different values of viscosity; air has a very low viscosity, water is more viscous, oil is even more so, and treacle is very viscous. The normal quantity symbol for viscosity is μ, and its units are $N\,s\,m^{-2}$ or $kg\,m^{-1}\,s^{-1}$.)

If we denote any one of the frequencies of vibration by ω, we can say that

$$\omega = f(l, E, G, \rho_m, v, \rho_a, \mu). \tag{15}$$

The next step is to replace each parameter by its dimensions

$$\left[\frac{1}{T}\right] = \phi \left[L, \frac{M}{LT^2}, \frac{M}{LT^2}, \frac{M}{L^3}, \frac{L}{T}, \frac{M}{L^3}, \frac{M}{LT}\right].$$

Unlike the case discussed in Section 8.2, here there is not a single unique combination of these groups which will yield a dimensionally homogeneous equation. We have to make use of the general rule stated at the end of that section.

We have a total of eight variables (including the dependent one ω), all of which we can express in terms of three basic dimensions $[M]$, $[L]$

and $[T]$. Thus, from the rule, our final relationship must be arranged to contain five (that is eight minus three) non-dimensional groups of variables.

The criteria for selecting our groups of variables are:
1 the resulting groups must be non-dimensional;
2 there must be five such groups;
3 each variable must appear in at least one of the five groups.

Looking at the variables and their associated dimensions, one non-dimensional group can be obtained by taking the ratio E/G, since both E and G have the same dimensions $[ML^{-1}T^{-2}]$. Similarly ρ_m/ρ_a forms another dimensionless group. The variables not yet included in groups are ω, l, v and μ.

Three of these can be combined into one dimensionless group $\omega l/v$ while the viscosity can be combined with ρ_a, v and l to form another non-dimensional group $\mu/\rho_a vl$.

SAQ 6
SAQ 6

Check that the groups $\omega l/v$ and $\mu/\rho_a vl$ are non-dimensional.

Thus we now have four non-dimensional groups in which all the variables feature. Nevertheless it is essential to choose yet another dimensionless group to make five. This could be

$$\left(\frac{G}{\rho_m v^2}\right), \qquad \left(\frac{\mu}{\rho_a l^2 \omega}\right), \qquad \left(\frac{E}{\rho_m \omega^2 l^2}\right), \qquad \text{etc.,}$$

selected arbitrarily. We can take any one of these to make up the final equation. (In practice, the one used will probably be governed by experimental considerations.)

Taking the first of the alternative groups $(G/\rho_m v^2)$, our final expression becomes

$$\left(\frac{\omega l}{v}\right) = f_1\left\{\left(\frac{E}{G}\right), \left(\frac{\rho_m}{\rho_a}\right), \left(\frac{\mu}{\rho_a vl}\right), \left(\frac{G}{\rho_m v^2}\right)\right\}. \tag{16}$$

This expression shows that each of the frequencies of vibration of the structure in question is directly proportional to v and inversely proportional to l as long as the dimensionless groups on the right-hand side are maintained constant in value.

In fact $\omega l/v$ is proportional to the product of power series in each of the four groups on the right-hand side of equation (16). To define the complete relationship it is necessary to determine each of the four power series. In principle this can be done most economically by conducting four series of experiments, in each of which all but one right-hand side group are kept constant in value. For example, in the first series of experiments E/G, ρ_m/ρ_a, and $\mu/\rho_a vl$ would be maintained constant while $G/\rho_m v^2$ was varied. The particular frequency of vibration of interest would be determined empirically so that values of $\omega l/v$ could be plotted against values of $G/\rho_m v^2$ and thus the polynomial in $G/\rho_m v^2$ determined. The procedure would then be repeated for the other dimensionless groups.

We begin an investigation of this kind (on the vibrations of plates) in the television programme for this unit and complete it in the broadcast notes.

SAQ 7
SAQ 7

Consider the two degree-of-freedom system shown in Figure 11. Obtain a set of dimensionless groups of variables which can be used to determine the displacement amplitude X_1 of the mass m_1 during steady-state forced vibrations.

8.4 An internal combustion engine problem

Let us now look at another situation where dimensional analysis can point the way to a solution of an otherwise complex problem. Suppose we wish to find out how the side thrust F_s on the piston of an internal combustion (IC) engine depends on the geometrical, dynamical, and operational parameters of the engine. This relationship is of interest where lubrication and engine wear are being studied.

The geometrical parameters of the engine are the crank and connecting-rod lengths r and l respectively, and the crank orientation θ at which the force is to be determined.

The dynamic parameters are the masses of the connecting-rod m_c and the piston m_p, the position of the centre of mass of the connecting-rod l_G and the rod's second moment of mass I.

Operational variables are the rotational speed ω and angular acceleration α of the crankshaft of the engine, the mean gas force F_g on the piston due to the burning fuel/air mixture, and the rate of change of gas force during combustion d_F, which is used as an indicator of the extent F_g changes during the cycle of events

Thus we write

$$F_s = f(r, l, \theta, m_c, m_p, l_G, I, \omega, \alpha, F_g, d_F)$$

which in dimensionalized form gives

$$[MLT^{-2}] = \phi[L, L, LL^{-1}, M, M, L, ML^2, T^{-1}, T^{-2}, MLT^{-2}, MLT^{-3}].$$

From twelve parameters we expect an equation involving nine non-dimensional groups. Self-evident non-dimensional groups which can be extracted from the list of variables are

$$\left(\frac{l}{r}\right), \quad \left(\frac{m_c}{m_p}\right), \quad \left(\frac{l_G}{l}\right), \quad \left(\frac{\alpha}{\omega^2}\right), \quad \theta.$$

Others which can fairly easily be deduced, and which you should check, are

$$\left(\frac{I}{m_c l^2}\right) \quad \text{and} \quad \left(\frac{F_g \omega}{d_F}\right).$$

So far we have seven dimensionless groups and the only variable not yet included is F_s. To incorporate this we could have, for example,

$$\left(\frac{F_s}{m_p \omega^2 r}\right), \quad \left(\frac{F_s}{m_c \omega^2 l}\right), \quad \text{or} \quad \left(\frac{F_s l}{I \alpha}\right).$$

Since, with any one of these, all the variables will then be incorporated, the ninth group can be chosen completely arbitrarily. It could be, for example,

$$\left(\frac{F_g}{m_p \omega^2 r}\right), \quad \left(\frac{F_g}{\alpha m_c l}\right) \quad \text{or} \quad \left(\frac{F_s}{F_g}\right).$$

So one form of the dependency is

$$\frac{F_s}{m_p \omega^2 r} = f_1\left\{\frac{l}{r}, \frac{m_c}{m_p}, \frac{l_G}{l}, \frac{\alpha}{\omega^2}, \theta, \frac{I}{m_c l^2}, \frac{F_g}{m_p \omega^2 r}, \frac{F_g \omega}{d_F}\right\}.$$

That is, the piston side load is proportional to $m_p \omega^2 r$ but the constant of proportionality is modified, not as each of the eleven dependent variables are changed, but only as each of the eight dimensionless ratios is changed. The actual value of this constant must be found by experiment.

We can use this example to demonstrate the type of result which can be obtained.

If we study the side thrust on the piston of an IC engine while it is being *motored* at constant speed (that is driven, without sparking plugs in, by an external power source), we find that the maximum value of the side thrust is virtually independent of crank angle. That is when α, F_g and d_f are all zero, F_s is not a function of θ. Therefore

$$\frac{F_s}{m_p\omega^2 r} = f_1\left(\frac{l}{r}, \frac{m_c}{m_p}, \frac{l_G}{l}, \frac{I}{m_c l^2}\right).$$

For a limited range of values of the variables relevant to small diesel engines, it can be shown by further analysis that the function is actually defined by

$$\frac{F_s}{m_p\omega^2 r} = -A_2\left(\frac{m_c}{m_p}+0.24\right)\left(\frac{I}{m_c l^2}+0.36\right)\left[\left(\frac{l_G}{l}\right)^2-0.83\frac{l_G}{l}-0.07\right]$$

$$\times\left[\left(\frac{l}{r}\right)^2-11.3\frac{l}{r}+84.4\right].$$

Despite its daunting detail, this equation confirms that $F/m_p\omega^2 r$ is still a function only of l/r, m_c/m_p, l_G/l and $I/m_c l^2$, and this is the information most relevant to the design of any series of model or prototype tests. (Note, however, that this does not apply to normal running, only to pressureless freewheeling.)

SAQ 8

The resistance experienced by a sea-going vessel propelling itself forward through the water comprises a frictional resistance due to the viscosity of the water and a wave-making resistance due to the fact that the water must be raised in a bow wave before it can accelerate and move apart as the vessel progresses. The resistance force can be considered to be a function of the speed of the vessel v, its length l, the gravitational acceleration g (this relates the mass of water displaced to its weight, which in turn affects the generation of water waves), the water density ρ and its viscosity μ. Determine a non-dimensional equation resulting from dimensional analysis of the problem.

8.5 Difficulties in application

The application of dimensional analysis appears to be extremely simple, but in an area which is not already well-trodden a number of difficulties may well arise.

1 The omission of an important variable from the list of independent variables is probably the most obvious example of an error which will lead to misleading results. The only safeguard is a thorough understanding of the physical nature of the problem to be solved and an appreciation, in a qualitative way, of the interrelation of factors affecting the situation. For example, if the acceleration α had been omitted from the list of variables for the piston side thrust, any resulting expression would be applicable only to the engine rotating at constant speed, and large errors could result if attempts were made to apply it to an accelerating engine.

2 There is always the possibility of including an unimportant variable, one whose effect on the situation is negligible. Examples are the effect of surface tension on the resistance to motion of ocean-going vessels and the effect of changes of ambient temperature (such as might occur at ground level in the UK) on natural frequencies of vibration of steel structures. The effect of including an unnecessary variable is merely to generate an additional non-dimensional group. This might cause real difficulty in the case of physical modelling or unnecessary work in the

case of empirical determination of a dependence. In either case it will be helpful to be able to make a qualitative assessment of the likely magnitude of the influence of the variable and hence decide whether or not to retain it. It is not unusual to get an inconvenient or misleading set of groups at the first attempt, but a combination of analysis, common sense and experience enables the errors to be spotted and corrected.

3 There is also the possibility of including a variable whose effect is already expressed fully by other variables. For example, Poisson's ratio v is fully covered by the inclusion of both the moduli of elasticity and of rigidity for any elastic material. Again an additional non-dimensional group will result. Although the duplication should become evident at a later stage, there may be modification and simplification of the original result of dimensional analysis so that the duplication is masked and remains undetected. This would give rise to confusion. It is therefore good practice to consider, at the earliest opportunity, whether there has been any duplication.

4 There is always the probability that a function cannot be defined for a sufficient range of interest of the variables, by an equation of the assumed form. For example, we might start off with

$$\omega = f(l, \rho, E, b, d)$$

and derive

$$\frac{\omega l}{\sqrt{(E/\rho)}} = f_1\left(\frac{b}{l}, \frac{d}{l}\right).$$

If we keep d/l constant and, as a result of a number of experiments, we plot values of $\omega l/\sqrt{(E/\rho)}$ against values of b/l, we can try and identify the resulting curve. If we have only a limited range of the variables we might think that the curve is a straight line and not recognize that what we have is a small section of, say, a trigonometric or exponential curve. Thus our result is strictly limited to the range of variables investigated. Although interpolation is reasonably safe, extrapolation can be very dangerous.

8.6 Physical significance of non-dimensional groups

Having set up an equation in terms of non-dimensional groups, it is helpful to have a physical interpretation of the 'meaning' of the constituent non-dimensional groups.

One of our first examples involved the natural frequencies of vibration of a simple beam, and the term $\omega l\sqrt{(\rho/E)}$. The description of this parameter is simplified if we consider the square of the term, $\omega^2 l^2 \rho/E$. The variable ρ is mass per unit volume and $\omega^2 l$, having dimensions of $[L/T^2]$, is akin to an acceleration. So $\omega^2 l \rho$ can be thought of as the acceleration force on a unit volume of material. The variable E is a representation of stiffness and has units of force per unit area. This can be conceived of as a force per unit volume by dividing by length to give E/l. So the ratio of these two,

$$\frac{\omega^2 l \rho}{(E/l)} = \frac{\omega^2 l^2 \rho}{E}$$

is a ratio of acceleration forces to stiffness forces, or more specifically, is the ratio of vibrational acceleration forces to bending forces in the system in question.

Similarly, the Reynolds and Froude numbers (introduced in SAQ 8) and the Mach number (which you will meet in Unit 12) are respectively the ratios of acceleration forces to viscous forces, to gravitational forces and to compressibility forces in flowing fluids. All these are dealt with in detail in Unit 12.

PHYSICAL MODELLING

We have already seen that the technique of dimensional analysis suggests a way in which physical scale models of, say, plant apparatus or structures can be used to predict the performance of the prototype. We now return to the problem of modelling and use equations derived by dimensional analysis to help us apply the ideas of similarity to more complicated situations. As long as the numerical values of non-dimensional parameter groups on the right-hand side of the appropriate equation can be kept the same for the model as for the prototype, then the numerical value of the left-hand side group (which must be determined empirically from the model), will be the same for both and can be used to predict some aspect of the performance of the true scale system.

Obviously, there must be some good reason for constructing a model in the first place rather than performing the requisite measurements on the prototype. One reason might be the size of the prototype. For example, in the case where information is needed for the design of a suspension bridge, the use of a small scale model is essential.

The first and most obvious requirement is that the model should be a geometrically scaled replica of the object under investigation. The universal acceptance of this need for geometric similarity is the reason why, in nearly all dimensional analyses, only a single characteristic length parameter is included. It is assumed that since the condition for geometric similarity will be observed, there is no need to include additional representation of the geometry of the system. Geometric similarity is therefore an essential requirement for the type of physical modelling under discussion. In addition we must have both kinematic similarity and dynamic similarity. Kinematic similarity is achieved when we ensure that there is a constancy of terms which relate velocities like $\omega l/v$ in equation (16). Dynamic similarity is similarly associated with terms which relate such variables as forces, elastic constants and densities.

To achieve dynamic similarity for our model suspension bridge, we must maintain all the right-hand non-dimensional groups of equation (16) constant. The requirement that E/G should be the same for the model and full scale structure probably (though not necessarily) means that we must use the same material for both. Maintaining the same value for $G/\rho_m v^2$ may then mean that the wind speed v must be the same in the two cases. The group $\mu/\rho_a vl$ must also be kept constant which means (if we assume that μ itself is almost constant) that the product $\rho_a l$ must be a constant. This implies that a scale model smaller than full size must be tested at a proportionally greater ambient air pressure. But this will violate the requirement for ρ_m/ρ_a to be constant and so we have a contradictory requirement which, strictly, we cannot solve.

We overcome this difficulty, however, by reassessing the relative importance of the several independent variables. It is probable that, in the case of a suspension bridge, the frictional drag due to the passing air is very small in comparison with the other forces. If μ is then deleted from the list of variables we have no conflict. Because now

$$\frac{\omega l}{v} = f_1 \left[\frac{E}{G}, \frac{\rho_m}{\rho_a}, \frac{G}{\rho_m v^2} \right],$$

all the right-hand dimensionless groups can be kept constant. The model is tested at the airspeed to which the full-size structure is subject. The natural frequencies of the model bridge can then be measured and, since

$\omega l/v$ is the same for prototype and model, the full scale bridge can be expected to vibrate at the same air speeds but at frequencies which are lower than those measured on the model by the magnitude of the geometric scale factor.

One limit to how small an effective scale model can be is set by the need to ensure that, in the model, the viscous forces due to the air motion do not become significant. This precludes the use of a very small model. A more restrictive limit might be set by the geometric scaling requirements. The cables of suspension bridges are multi-stranded and, in trying to produce a small scale version of this it is almost certain that the smaller strands will have a greater relative strength. Even if it is decided that the real cables can be adequately modelled by a single unstranded wire, there will be some other detail which cannot be adequately modelled if the scale is very small. Typical examples of features which set a lower limit to the scale factor are pre-stressing and re-inforcing of concrete, riveted and spot-welded joints in structures, and the need for scaling of clearances in bearings (these distances are so small that they may be affected by the surface finish of the model).

Section 10

SUMMARY

1   In mechanics, the units of all quantities can be expressed in terms of the three basic dimensions of length, time and mass.

2   Objects or systems which are geometrically similar, having all their corresponding dimensions in the same ratio, may look similar but not exhibit similar behaviour.

3   Two systems may be kinematically similar as well as geometrically similar, in which case corresponding points must have velocities which are in a constant ratio. This condition relates not only to the length dimension but also to the time dimension.

4   For two systems to be dynamically similar they must additionally have similar force distributions and their respective forces must be in a constant ratio (which relates to the dimension of mass). When this requirement is satisfied the systems are also geometrically and kinematically similar, and are said to be completely similar.

5   The technique of dimensional analysis consists of listing the dependent variables for one independent variable, noting their dimensions (usually in terms of $[M]$, $[L]$ and $[T]$) and arranging the variables into dimensionless groups. This reduces the degree of uncertainty in the form of the relationship, thereby reducing the number of independent sets of experiments which would have to be performed to determine the relationship.

6   The technique of dimensional analysis itself is simple. However, before doing the analysis one must compile a list of variables, and this poses the difficult problem of which to omit rather than which to include. After doing the analysis one usually has to decide how to construct a physical model in the face of impossible simultaneous conditions for dynamic similarity. These two operations are intriguing and challenging, making dimensional analysis more an art than a science. Expertise can be developed only by experience, and unfortunately we can provide only a little of this for you in this course.

Not Examinable.

SOLUTION OF THE FORCED DAMPED VIBRATION EQUATION

An alternative method for obtaining the steady-state response derived in Section 3 is by the use of a vector diagram.

We begin, as before, with equation (4)

$$m\ddot{x} + c\dot{x} + kx = F \sin \omega t,$$

and we assume the same form for the steady-state solution as that given in equation (5),

$$x = D \sin \omega t + E \cos \omega t = X \sin (\omega t \pm \phi)$$

where ϕ could be either negative or positive and $X = \sqrt{(D^2 + E^2)}$, the amplitude of motion of the mass.

It follows that

$$\dot{x} = \omega X \cos (\omega t \pm \phi) = \omega X \sin \left(\omega t \pm \phi + \frac{\pi}{2} \right).$$

This means that the vector which represents the velocity of the mass is 90° ahead of the vector representing displacement. Similarly,

$$\ddot{x} = -\omega^2 X \sin (\omega t \pm \phi) = \omega^2 X \sin (\omega t \pm \phi + \pi).$$

Hence the vector of the acceleration of the mass is parallel to the displacement vector but has the opposite arrowhead.

Until we know the sign of ϕ, let us put $(\omega t \pm \phi) = \theta$.

The last term on the left-hand side of equation (4) then becomes

$$kx = kX \sin \theta.$$

Suppose we specify an arbitrary reference direction. Then $kX \sin \theta$ represents the component at right angles to this direction of a vector of length kX which is inclined at an angle θ to that direction (Figure 17).

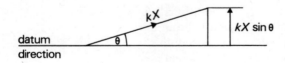

Figure 17

A similar argument applies to all the other terms in equation (4), using the appropriate lengths and angles. Thus we can represent equation (4) by the vector diagram shown in Figure 18. This diagram leads to the following conclusions:

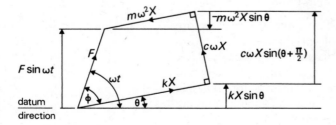

Figure 18   Vector diagram representing the forced damped vibration equation

52

1

$$\theta = (\omega t - \phi)$$

so that ϕ is negative, that is the displacement lags behind the external force.

2

$$\tan \phi = \frac{c\omega X}{kX - m\omega^2 X} = \frac{c\omega}{k - m\omega^2}$$

that is, the same magnitude as $\tan \phi$ derived in Section 3, the sign having been determined in 1 above.

3   When $m\omega^2 X > kX$, that is when $\omega^2 > k/m$, then $\phi > 90°$.

4

$$F^2 = (c\omega X)^2 + (kX - m\omega^2 X)^2 = X^2[c^2\omega^2 + (k - m\omega^2)]^2,$$

therefore

$$X = \frac{F}{\sqrt{[(k - m\omega^2)^2 + (c\omega)^2]}}$$

as before.

A RULE FOR DIMENSIONAL ANALYSIS

Here we give an example which illustrates the rule of dimensional analysis quoted at the end of Section 8.2. It is not a proof, but will help give you an intuitive understanding of the rule.

We take the example of the deformation of motor cars as a result of collision. The extent of the impact deformation can be characterized by the depth of penetration (x). This will be a function of the velocity and acceleration at the moment of impact (v and a respectively), the size of the car (l), and the thickness of the material of which the car is made (b). In addition it will vary with the strength and weight of the structural material of the car. These are characterized by the elastic modulus (E), the yield stress (Y) and the density (ρ). Thus

$$x = f(v, a, l, b, E, Y, \rho) \qquad (A1)$$

by which we mean that x is some explicit function (as yet unknown) of v, a, l, b, E, Y and ρ. Let us assume that this function is the simple product of unknown powers of each of the variables. That is, that

$$x^{\mu} = k \times v^c \times a^d \times l^h \times b^j \times E^n \times Y^r \times \rho^s \qquad (A2)$$

where μ, k, c, d, h, j, n, r and s, are all purely numerical quantities. If we replace the symbols for the variables by the appropriate dimensions we get

$$[L^{\mu}] = \left[\frac{L}{T}\right]^c \left[\frac{L}{T^2}\right]^d [L]^h [L]^j \left[\frac{M}{LT^2}\right]^n \left[\frac{M}{LT^2}\right]^r \left[\frac{M}{L^3}\right]^s.$$

For the equation to be dimensionally homogeneous, the sum of the powers of $[M]$ on the left-hand side of the equation must equal the sum of the powers of $[M]$ on the right-hand side, and similarly for $[L]$ and $[T]$.

For $[M]$: $\qquad 0 = n + r + s.$ $\qquad\qquad$ (A3)

For $[L]$: $\qquad \mu = c + d + h + j - n - r - 3s.$ $\qquad$ (A4)

For $[T]$: $\qquad 0 = -c - 2d - 2n - 2r.$ $\qquad\qquad$ (A5)

Thus we have three equations, one corresponding to each of our fundamental units, in terms of eight unknowns, one corresponding to each of the variables in the equation (A1). But with only three equations we can solve for only three unknowns. The particular three for which we solve can be chosen quite arbitrarily. For example if we choose to solve for c, h and n:

from (A3) $\qquad n = -r - s;$

from (A5) $\qquad c = -2d - 2n - 2r$

$\qquad\qquad\qquad = -2d + 2s;$

from (A4) $\qquad h = -c - d - j + n + r + 3s + \mu.$

Therefore equation (A2) becomes

$$x^{\mu} = k\left(v^{(-2d+2s)} a^d l^{(-c-d-j+n+r+3s+\mu)} b^j E^{(-r-s)} Y^r \rho^s\right)$$

which can be written in dimensionless form as

$$\left(\frac{x}{l}\right)^{\mu} = k\left(\frac{al}{v^2}\right)^d \left(\frac{b}{l}\right)^j \left(\frac{Y}{E}\right)^r \left(\frac{\rho v^2}{E}\right)^s (l)^{(-c-2d+n+r+3s)} \qquad (A6)$$

or, since the index of the l term in equation (A6) is zero (check this by substituting for n and c from (A3) and (A5)), it can be written

$$\frac{x}{l} = f_1\left(\frac{al}{v^2}, \frac{b}{l}, \frac{Y}{E}, \frac{\rho v^2}{E}\right). \tag{A7}$$

In equation (A7) there are five dimensionless groups, one corresponding to each of the indeterminate powers μ, d, j, r and s in equation (A2). This thus confirms the rule that the equation resulting from any dimensional analysis will contain a number of non-dimensional groups, this number being equal to the initial number of variables minus the number of fundamental dimensions embodied in the equation. (You can see in our example that the number of fundamental dimensions was equal to the number of indices for which we could solve.)

This statement can be re-expressed in two more rigorous theorems, which vary slightly from it in their complete form. However, this version of the rule is adequate for our purposes in this course.

Suppose now that instead of solving equations (A3), (A4) and (A5) for c, h, and n, we had solved for d, j and s. We would have obtained the expression

$$\frac{x}{b} = f_1\left(\frac{v}{\sqrt{(ab)}}, \frac{l}{b}, \frac{E}{\rho ab}, \frac{Y}{\rho ab}\right). \tag{A8}$$

The term $v/\sqrt{(ab)}$ in equation (A8) can be obtained by taking the square-root of the reciprocal of the product of the first two terms in the right-hand side function of equation (A7). The term $E/\rho ab$ can be obtained by taking the reciprocal of the product of all terms except Y/E in the function of equation (A7). This illustrates the general rule that any particular form of the equation resulting from dimensional analysis (such as (A7) or (A8)) can be modified at will, by multiplying any non-dimensional group by any power (positive or negative) of any other non-dimensional group. The resulting equation will be just as valid as the initial equation, as long as it contains the requisite number of non-dimensional groups and as long as all the original variables are still represented in the final equation.

This result reflects the fact that it cannot be possible to determine completely the form of an equation by knowing only the dimensions of the constituent variables. It justifies the way that non-dimensional groups were selected arbitrarily from combinations of the variables in equation (A2). In the general case, the only criteria are that in the final equation there must be the correct number of non-dimensional groups and that all variables must appear at least once. After that the task becomes one of finding the most convenient way to discover useful details about the function.

ANSWERS TO SELF-ASSESSMENT QUESTIONS

SAQ 1

From equation (7), the steady-state amplitude of the mass is given by

$$X = \frac{F}{\sqrt{[(k - m\omega^2)^2 + (c\omega)^2]}}.$$

When $\omega^2 =$ (natural undamped angular frequency)^2, we have

$$\omega^2 = \omega_n^2 = \frac{k}{m} = \frac{15}{0.1} = 150 \ (\text{rad s}^{-1})^2$$

and since $X = 0.02$ m, when the damping coefficient has its least acceptable value,

$$(0.02)^2 = \frac{(0.7)^2}{(15 - 15)^2 + 150c^2}$$

$$150c^2 = \frac{(0.7)^2}{(0.02)^2},$$

$$c^2 = \frac{(0.7)^2}{150 \times (0.02)^2} = 8.17.$$

Thus the least damping coefficient required is

$$c = 2.86 \ \text{N s m}^{-1}.$$

The corresponding phase angle is given by

$$\tan\phi = \frac{-c\omega}{k - m\omega^2} = -\frac{2.86 \times \sqrt{150}}{15 - 15} = -\frac{2.86\sqrt{150}}{0}$$

hence $\tan\phi$ is infinitely large so that $\phi = -90°$. Thus the displacement lags the applied force by 90°. Note that this value of ϕ is the same for all values of c when $k = m\omega^2$, that is when the forcing frequency equals the natural undamped frequency.

SAQ 2

Still using curve 1 of Figure 6(a) (the damping ratio α/ω_n still equals 0.1), we find that for $Xk/F = 1.2$, $\omega/\omega_n = 1.34$. Therefore

$$\omega_n = \frac{\omega}{1.34} = \frac{60\pi}{1.34} = \sqrt{\frac{k}{150}}.$$

Hence the stiffness is

$$k = \left(\frac{60\pi}{1.34}\right)^2 \times 150 = 2.97 \times 10^6 \ \text{N m}^{-1}$$

$$= 2.97 \ \text{MN m}^{-1}.$$

Since $\alpha/\omega_n = 0.1$,

$$\alpha = 0.1 \times \omega_n = 0.1 \times \frac{60\pi}{1.34} = 14.07,$$

and since $\alpha = c/2m$, the new damping coefficient is

$$c = 14.07 \times 300 = 4221 \ \text{N s m}^{-1} \quad \text{or} \quad 4.221 \ \text{kN s m}^{-1}.$$

From curve 1 in Figure 6(b), for $\omega/\omega_n = 1.34$,

$$\phi = 2.76 \ \text{rad} = 158.1°.$$

SAQ 3

(a) From the curve in Figure 6(a) for a damping ratio α/ω_n of 0.25,

resonant frequency = 0.95 × natural undamped frequency

that is

$$\omega_n = \frac{30}{0.95} = 31.6 \ \text{rad s}^{-1}$$

therefore

$$\frac{k}{m} = (31.6)^2, \quad k = 0.1 \times 1000 = 100 \ \text{N m}^{-1}.$$

Also from the graph,

$$\left(\frac{kX}{F}\right)_{\text{max}} = 2.075.$$

Therefore

$$X_{\text{max}} = 2.075 \times \frac{2}{100} = .0415 \ \text{m}.$$

(b) The required range of values of ω is $0.3\omega_n$ to $1.28\omega_n$, that is $9.5 \ \text{rad s}^{-1}$ to $40.5 \ \text{rad s}^{-1}$.

(c) From the graph of Figure 6(b),

phase angle at $0.3\omega_n = 0.18$ rad or 10.3°;

phase angle at $1.28\omega_n = 2.36$ rad or 135.2°.

SAQ 4

(a) We know that $\omega_n = \sqrt{(k/M)}$ where M is the mass of the machine.

When the load on the springs is Mg, the deflection is 0.002 m (the dashpot is not effective at zero velocity.) Hence the stiffness

$$k = \frac{Mg}{0.002} = 500Mg \ \text{N m}^{-1}.$$

Hence

$$\omega_n = \sqrt{\frac{k}{M}} = \sqrt{(500g)}$$

$$= \sqrt{(500 \times 9.81)} = 70.04 \ \text{rad s}^{-1}.$$

(b) When $\omega = 61.3 \ \text{rad s}^{-1}$,

$$\frac{\omega}{\omega_n} = \frac{61.3}{70.04} = 0.875.$$

When $\omega = 107.6 \ \text{rad s}^{-1}$,

$$\frac{\omega}{\omega_n} = \frac{107.6}{70.04} = 1.536.$$

On Figure 9, the curve which gives the same value of $Ym/\epsilon m_e$ for both these values is curve 2 hence the required damping ratio

$$\frac{\alpha}{\omega_n} = 0.25.$$

(c) From curve 2, the value of $Ym/\epsilon m_e$ at the given values of ω/ω_n is 1.5. We also know that $Y = 0.9$ mm and $m/m_e = 20$, therefore

$$1.5 = \frac{0.9 \times 20}{\epsilon},$$

$$\epsilon = \frac{0.9 \times 20}{1.5} = 12 \ \text{mm}.$$

(d)

$$\frac{\omega}{\omega_n} = \frac{107.6}{70.04} = 1.536 \quad \text{and} \quad \frac{\alpha}{\omega_n} = 0.25.$$

Hence from curve 2 on Figure 6(b)

$$\phi = 2.62 \ \text{rad} = 150.1°.$$

SAQ 5

Let F be the braking force and a the constant deceleration. Then

$$\frac{F_m}{F_p} = \frac{1}{50} = \frac{m_m a_m}{m_p a_p}.$$

We also know that

$$\frac{m_m}{m_p} = \frac{\rho l_m^3}{\rho l_p^3} = \left(\frac{l_m}{l_p}\right)^3 = \frac{1}{1000}.$$

Therefore the acceleration ratio

$$\frac{a_{\mathrm{m}}}{a_{\mathrm{p}}} = \frac{1}{50}\bigg/\frac{m_{\mathrm{m}}}{m_{\mathrm{p}}}$$

$$= \frac{1000}{50} = 20.$$

But

$$\frac{a_{\mathrm{m}}}{a_{\mathrm{p}}} = \frac{l_{\mathrm{m}}}{l_{\mathrm{p}}}\bigg/\left(\frac{t_{\mathrm{m}}}{t_{\mathrm{p}}}\right)^2$$

therefore

$$\left(\frac{t_{\mathrm{m}}}{t_{\mathrm{p}}}\right)^2 = \frac{l_{\mathrm{m}}}{l_{\mathrm{p}}}\bigg/\frac{a_{\mathrm{m}}}{a_{\mathrm{p}}} = \frac{1}{10 \times 20}.$$

Hence the ratio of stopping times is

$$\frac{t_{\mathrm{m}}}{t_{\mathrm{p}}} = \frac{\sqrt{200}}{200} = \frac{14.14}{200} = 0.0707,$$

and a time interval of 7s for the model corresponds to a time interval of 100 s in the prototype.

The ratio of initial velocities is therefore

$$\frac{u_{\mathrm{m}}}{u_{\mathrm{p}}} = \frac{l_{\mathrm{m}}}{l_{\mathrm{p}}}\bigg/\frac{t_{\mathrm{m}}}{t_{\mathrm{p}}} = \frac{200}{10\sqrt{200}} = \frac{\sqrt{200}}{10} = 1.414.$$

For a constant deceleration from an initial velocity u to a final velocity of zero, $u^2 = 2as$, where s is the stopping distance. Thus

$$\frac{s_{\mathrm{m}}}{s_{\mathrm{p}}} = \left(\frac{u_{\mathrm{m}}}{u_{\mathrm{p}}}\right)^2\bigg/\left(\frac{a_{\mathrm{m}}}{a_{\mathrm{p}}}\right) = \frac{200}{100 \times 20} = \frac{1}{10},$$

that is a stopping distance of 10 m in the model corresponds to a stopping distance of 100 m in the prototype.

SAQ 6

$\dfrac{\omega l}{v}$ has dimensions of $\dfrac{[1/T][L]}{[L/T]}$ which is dimensionless.

$\dfrac{\mu}{\rho_a v l}$ has dimensions of

$$\frac{[M/LT]}{[M/L^3][L/T][L]} = \left[\frac{M}{LT}\right]\left[\frac{L^3}{M}\right]\left[\frac{T}{L}\right]\left[\frac{1}{L}\right] \quad \text{which is dimensionless.}$$

SAQ 7

We can write

$$X_1 = f(m_1, m_2, k_1, k_2, c_1, c_2, F, \omega),$$

giving

$$[L] = \phi\left[M, M, \frac{M}{T^2}, \frac{M}{T^2}, \frac{M}{T}, \frac{M}{T}, \frac{ML}{T^2}, \frac{1}{T}\right].$$

We need six dimensionless groups, for example

$$\frac{m_1}{m_2}, \frac{k_1}{k_2}, \frac{c_1}{c_2}, \frac{k_1}{m_1\omega^2}, \frac{c_2\omega}{k_2}, \frac{F}{k_1 X_1}.$$

Hence we have

$$\frac{X_1 k_1}{F} = f_1\left(\frac{m_1}{m_2}, \frac{k_1}{k_2}, \frac{c_1}{c_2}, \frac{k_1}{m\omega^2}, \frac{c_2\omega}{k_2}\right).$$

(Note that there is no need to include X_2 in the variables because once F, ω and the parameters of the system are specified, the steady-state values of X_1 and X_2 are both determined.)

SAQ 8

We can write

$$\text{resistive force} = F = f(v, l, g, \rho, \mu),$$

giving

$$\left[\frac{ML}{T^2}\right] = \phi\left[\frac{L}{T}, L, \frac{L}{T^2}, \frac{M}{L^3}, \frac{M}{LT}\right].$$

So we expect three dimensionless groups. Starting with the ratio of μ to ρ (to eliminate M), we need to divide further by parameters having combined dimensions of $[L^2/T]$. The product vl satisfies this specification. Thus one dimensionless group is $\mu/\rho vl$. The resistive force can be non-dimensionalized with $\rho v^2 l^2$. The only parameter not then included is g, so this must feature in the last group which could be v^2/lg. Thus we have

$$\frac{F}{\rho v^2 l^2} = f_1\left(\frac{v^2}{lg}, \frac{\rho vl}{\mu}\right)$$

The two groups on the right-hand side of this equation are important in fluid mechanics, and are known as the Froude and Reynolds numbers respectively (further reference is made to them in Unit 12). The Froude number is usually used in the square root of this form, $v/\sqrt{(lg)}$.

57

Unit 9
Examples in Machine Design

CONTENTS

AIMS

This unit aims to put the material of previous units together in the context of simple design calculations, in order to demonstrate the interdependence and the practical importance of the various topics.

OBJECTIVES

After studying this unit you should be able:

1 To define the term *dynamic force*.

2 To use your knowledge of kinematics, statics and rigid body dynamics to determine dynamic forces in machine components which can be modelled as rigid bodies.

3 Given relevant dimensions of a machine component, to work out the stresses due to dynamic as well as static forces in simple cases.

4 To understand the difference between reasonable approximations and theoretically perfect calculations and the great importance of the former in engineering practice.

5 To define the terms *gas force*, *side thrust*, and *balancing mass* as applied to a reciprocating internal combustion (IC) engine.

6 Given the gas force, and the masses and main dimensions of the moving parts, to work out the output torque due to one cylinder of a reciprocating IC engine.

7 To understand the terms *cam*, *follower*, *point follower*, *roller follower*, *stroke*, *dwell*, and *pitch line*.

8 To identify dynamically undesirable features in a proposed follower motion.

9 Given the stroke, and the times of movement and dwell of the follower, to plot a cycloidal or sinusoidal displacement diagram for the follower.

10 To determine the required stiffness of the return spring for a follower performing a given cycloidal or sinusoidal motion.

STUDY GUIDE

The television programme 'Designing for power' and the home experiment on cams are associated with this unit. A disc discusses the exercise at the end of Section 3.2.4.

The appendix to this unit is optional: you will not be examined on it, but you should be able to *use* the theorem proved in it. It provides useful revision of some of the course material, therefore you should try to make time to work through it.

INTRODUCTION

Each previous unit of this course has concerned a particular aspect of engineering mechanics. The advantage of splitting the subject up like this is that each section can be covered thoroughly: the theory can be explained and the inherent limitations and assumptions pointed out. Exercises and SAQs can be used to reinforce the important points made in the text.

But it would not do to leave it at that. Teaching all the parts, however successfully, is not an adequate method of teaching the subject of engineering mechanics. The theory is based on a small number of simple fundamental principles (which means that all parts of the theory are closely related), but it can help to solve a wide range of practical problems. In order to teach the subject satisfactorily, therefore, it is necessary to demonstrate its use in coping with more complex practical problems where all available knowledge, experience and judgement are deployed according to the demands of the particular situation.

Although we are only about half way through our account of the various aspects of engineering mechanics, we have covered enough to enable us to look at some suitable problems in engineering design, which embraces all aspects of engineering theory and practice. We shall go through some typical design calculations in order to demonstrate some of the problems and the way in which a knowledge of mechanics can help to solve them. In other words we shall look at the way in which calculations can be used to ensure that the components of real engineering products are fit for their purpose.

It follows that this unit will contain little fundamentally new material; it will concentrate on applying the material covered by the previous units and on bringing together a number of topics which have so far been dealt with separately.

It will not be possible for us to cover every aspect of the design process, or to go through the complete design of a machine. Many of the most important features of such a procedure will have to be taken for granted in this course: for example the process of creative thinking or invention, and social and environmental questions. (To what extent is there a need for our proposed product? Who wants it? Whom will it benefit? What will be the effect on the surroundings of its installation and use?) We will not consider the use and availability of raw materials, fuel, labour and other essential resources, nor shall we say anything about the financial implications of design.

The unit presents three almost independent sections (Sections 2, 3 and 4). In the first we will examine the excavator, in the second the internal combustion engine and in the third, the design of cams.

DYNAMIC FORCES IN THE EXCAVATOR

We begin by introducing the important concept of **dynamic forces**. These are forces which are due to the need to accelerate machine components. Since such components rarely move with uniform velocity or remain stationary, dynamic forces are very common in machine design. In some machines they are the most important source of load and stress.

dynamic forces

In Unit 4 we did some 'static' calculations of the forces which arise when the excavator is stationary; we will now extend these calculations to take into account the effect of the motion of the various parts.

This excavator is a very versatile machine and can move in many different ways: here we shall look at a number of these, so that we can estimate the importance of the dynamic effects in each case. In order to find the relevant accelerations, we shall make use of the graphical techniques that were introduced in Unit 3. Our object will be to check that the components of the machine are strong enough to resist the dynamic loads that arise during its motion.

In Section 2.1 we shall look at the motions of the rams and the dynamic forces on the loaded bucket. In Section 2.2 we estimate the greatest angular acceleration which the dipper arm is likely to have and the consequent effect on its supports. We go on in Section 2.3 to consider the stresses in the dipper arm itself. Section 2.4 is concerned with the motion of the excavator as a whole.

2.1 A normal lifting operation

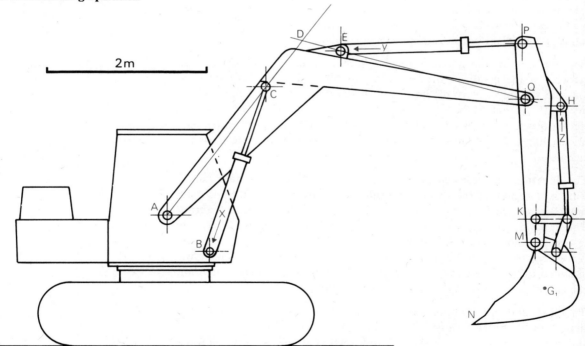

2m

Figure 1 The excavator: scale 1/48 full size

In our first investigation, the excavator is in the position shown in Figure 1: it is lifting the full bucket by using the twin hydraulic boom rams or cylinders to raise the boom, and by using the dipper-arm ram to rotate the dipper arm to which the bucket is attached.

The bucket is about to be emptied of its contents, but we shall assume that it is still full at the instant we are considering. We shall also assume a steady rate of oil flow into the hydraulic cylinders, which means that the piston rods are moving with a constant velocity. First we must arrive at an estimate of what these velocities might be.

The specification of the machine tells us that the diameter of each hydraulic boom cylinder is very nearly 0.1 m and that the maximum available rate of oil flow is about 132 litres per minute. Remember that 1 litre = 1000 $cm^3 = 1000 \times 10^{-6}$ $m^3 = 0.001$ m^3, so that 132 litre per minute = 0.132 m^3 min^{-1}. The cross-sectional area of each piston is

$$A = \frac{\pi}{4} \times (0.1)^2 = \frac{\pi}{4} \times 0.01 \text{ m}^2.$$

Each cylinder will take half the total oil flow and if each piston moves a distance x along the cylinder, the change in volume will be Ax (Figure 2). The movement of the piston along the cylinder must therefore be such that

$$A \frac{dx}{dt} = \text{rate of change of volume behind the piston}$$

$$= \text{flow rate of oil into the cylinder}$$

where dx/dt is the piston velocity.

Putting in the figures for our cylinder we get

$$\frac{\pi}{4} \times 0.01 \times \frac{dx}{dt} = \frac{1}{2} \times 0.132 \times \frac{1}{60}$$

$$\frac{dx}{dt} = \frac{0.132}{120} \times \frac{4}{\pi} \times \frac{1}{0.01}$$

$$= 0.14 \text{ m s}^{-1}.$$

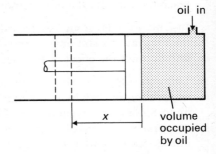

Figure 2   A hydraulic cylinder and piston

This is a maximum figure corresponding to the maximum available flow rate. If a similar calculation is done for the dipper cylinder (assuming the bucket cylinder to be locked and not requiring a supply of oil), its maximum piston velocity turns out to have a similar value.

For the velocity and acceleration diagrams in Figure 57 of Unit 3, the velocity of the piston in the boom-lift cylinder was taken as 0.04 m s^{-1} and the sliding velocity of the piston in the dipper-arm cylinder as 0.05 m s^{-1}; the piston of the dipper-arm cylinder was also assumed to have a sliding acceleration of 0.01 m s^{-2}. Notice that the sliding velocities used in the diagrams are about one third of the maximum that we have just calculated.

The directions of these velocities and accelerations are shown in Figure 3. This is a repeat of the diagrams from Unit 3, with the addition of the points g_1 on the velocity diagram and g_1' on the acceleration diagram. These points correspond to the velocity and acceleration of G_1, the (estimated) position of the centre of mass of the bucket shown in the displacement or configuration diagram (Figure 1). The bucket cylinder was assumed to be locked when the diagrams were drawn, and the tracks and the cab of the digger were assumed to be stationary.

The first and most important thing to notice about the acceleration diagram is that all the values of acceleration are very small compared with the standard free-fall acceleration. It follows that all the acceleration forces (that is the *dynamic* forces) are very small compared with the weights of the components. (Remember that a mass M at a point where the free-fall acceleration is g will have a weight of Mg; if the centre of mass is given an absolute acceleration a the accelerating force is Ma.)

Suppose, for example, that we look at the load on the dipper arm due to the bucket. The acceleration of the centre of mass of the bucket is about

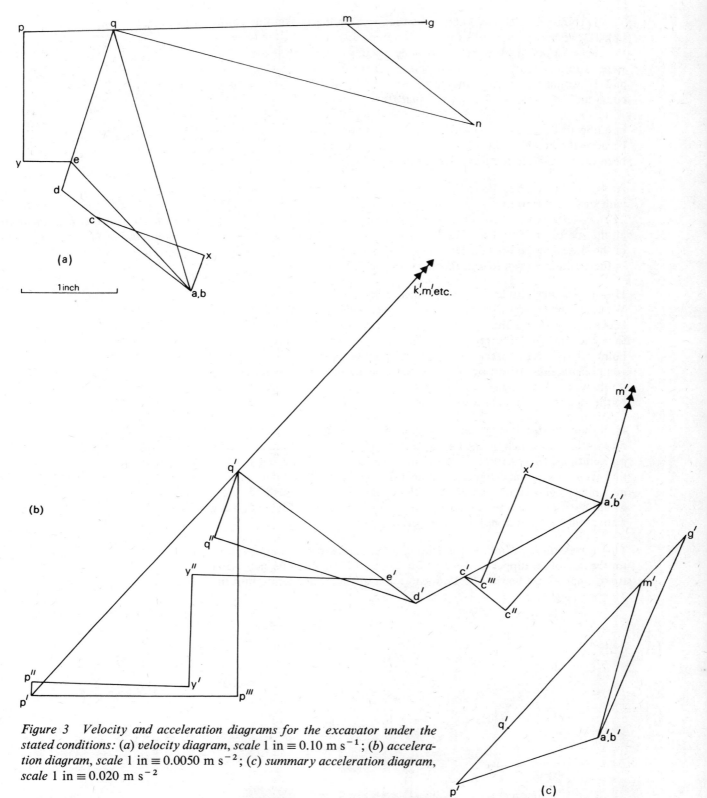

Figure 3 Velocity and acceleration diagrams for the excavator under the stated conditions: (a) velocity diagram, scale 1 in ≡ 0.10 m s⁻¹; *(b) acceleration diagram, scale* 1 in ≡ 0.0050 m s⁻²; *(c) summary acceleration diagram, scale* 1 in ≡ 0.020 m s⁻²

0.047 m s⁻² (the vector a′g′ on the acceleration diagram). The bucket itself has a mass of about 400 kg and it can carry about 1000 kg of earth, so that the total mass of the loaded bucket is about 1400 kg. Its weight, therefore, is about 13 750 N, and this weight is the static force which has to be supported by the dipper arm even when every part of the machine is at rest. The additional force due to the motion, that is the dynamic force, is equal to 1400 × 0.047 = 65.8 N, which is very small compared with the weight so that the vector sum of the static and dynamic forces between the bucket and the dipper arm is little different from the static force alone.

7

This is shown in Figure 4 where F_1 is the force on the bucket due to the link JL (we may assume this force to act along JL) and F_2 is the force on the bucket at pin M (its direction must be roughly as shown). If the bucket were at rest or moving with a constant velocity the vector sum of F_1, F_2 and W would equal zero: the vector sum of F_1 and F_2 would then be equal and opposite to W (that is equal to W in magnitude and directed vertically upwards). Figure 4 shows that the actual state of affairs corresponding to the motion we have specified is not very different from this because the accelerating force Ma is so small (its magnitude has in fact been exaggerated in the force diagram).

At first sight this may be surprising: one might expect that in a machine with such large and heavy links the dynamic forces would be considerable. The reason why they are not is quite simple: the links move very slowly; all the speeds are very low. The velocity of point G_1, for example, is one of the largest in the whole assembly, yet even that is only about $0.34\,\mathrm{m\,s^{-1}}$ in the position shown (check this from the velocity diagram).

How do the magnitudes of the velocities affect those of the accelerations? We know the answer to that, too. If the magnitude of the relative velocity between any two points a distance L apart on the same rigid link is v, then the normal (or centripetal) component of the acceleration of one point relative to the other is v^2/L. And since the tangential and Coriolis components must be drawn to the same scale as the normal components, it follows that the magnitudes of the relative accelerations are proportional to the squares of the relative velocities.

If all the velocities in Figure 3 were trebled to bring them up to the maximum velocities which, according to our calculations, could occur, all the accelerations derived from them would be multiplied by nine. This means that even if the boom cylinders and the dipper cylinder all had their pistons moving at their maximum velocities, the dynamic accelerations and forces would be only about nine times the values we obtain from Figure 3. They would still be small in normal engineering terms.

This leaves the question of the sliding acceleration which was specified for the ram of the dipper-arm cylinder in Figure 3. Could that be increased? If so, what would be the effect? What about a similar acceleration in the

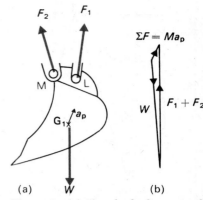

Figure 4 (a) *Free body diagram of the excavator bucket.* (b) *Force on the bucket*

A vivid example of pessimistic design: this ship had screws, paddles and sails! It is the Great Eastern Steamship, designed by Brunel and built over the period 1854–1859. (*Brunel University Library*)

boom cylinders? These are points well worth looking at and we deal with some of them in the next section by looking at an extreme case. In doing so we follow a useful principle in design: if in doubt, make your assumptions pessimistic; the actual performance must then be on the safe side of your predictions.

2.2 Jerking the dipper arm

To look at the effect of an acceleration of the piston in the dipper-arm cylinder, we shall assume the cab to be stationary and the boom cylinders to be locked. This will simplify the work and make little difference to the calculation, since as we have seen, the steady motion of the boom cylinders produces only small accelerations.

We shall also assume that the bucket is full, and that the bucket cylinder is locked so that the dipper arm and the bucket with its links and cylinder form one rigid body. We shall want to compare the static force on the pin at Q (Figure 1) with the force that it must exert when the dipper-arm ram has its greatest acceleration.

2.2.1 Static forces

Before we investigate the accelerated motion of the dipper-arm assembly, let us first look at the forces which act when this whole assembly is stationary. Most of these are gravity forces, and the weights of the various components are given in Table 1.

Table 1   Weights of the front end equipment components of the excavator

Component	Weight/N	Mass/kg	Estimated position of centre of mass in Figure 5
Dipper arm	2942	300	G_2
Bucket (full)	13750	1400	G_1
Bucket cylinder	981	100	G_3
Bucket links	490	50	G_4

(Some of these figures were given in Table 1 of Unit 4; notice that the weight of the bucket is well within the safe limit worked out in Unit 4 from the point of view of preventing the whole machine from toppling over).

It follows from these figures that the weight of the whole assembly is 18 163 N. (This neglects the unknown proportion of the weight of the dipper-arm ram which is supported on pin P, but this is not likely to be significant.)

To complete the static force analysis we need to know the position of the centre of gravity of the assembly; since we cannot calculate this precisely we shall have to be satisfied with a reasonable estimate. (As we have said before, precision is rarely attainable in engineering calculations; a good estimate of what can be expected is what is sought.) The position of the centre of gravity (which, remember, is at the same point as the centre of mass) will also be relevant to the analysis of the dynamic forces which we shall carry out later.

For the moment we mark reasonable estimates of the centre of gravity of each component on the scale diagram shown in Figure 5; the nomenclature used is that shown in the last column of Table 1. In dealing with the dipper arm itself, we shall assume it to be equivalent to a link of uniform width as shown by the dotted lines in Figure 5. This applies both now and later when we come to work out its second moment of mass. The centre of mass G_2 is positioned accordingly. Point G_4 will be midway between the centres of the two links of the bucket-tipping mechanism. G_3 is on the longitudinal axis of the bucket-tipping ram about half way between points J and H. The centre of mass of the bucket G_1 was used in the last section.

We now go on to find the position of the centre of mass of the whole assembly, assuming it to lie in the plane of the paper. The co-ordinates of the centre of mass in this plane can conveniently be defined relative to the perpendicular axes QG_1 and QO which define the central plane of the assembly.

We begin by finding the total first moment of mass about axis QG_1: we measure from Figure 5 the perpendicular distance of each component centre of mass from QG_1 and hence calculate individual values for the components, as shown in Table 2.

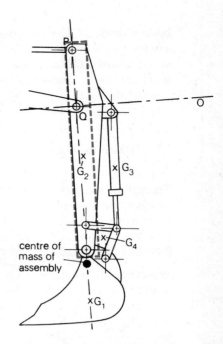

Figure 5 Front end equipment components of the excavator: scale 1/48 full size

Table 2 First moments of mass for front end equipment components about axis QG_1

Component	Mass/kg	Perpendicular distance of the centre of mass to axis QG_1/mm (from Figure 5)	First moment of mass about axis QG_1/kg mm
Dipper arm	300	1	300
Bucket (full)	1400	0	0
Bucket cylinder	100	9	900
Bucket links	50	4	200
		Total first moment of mass about QG_1 =	1400

The perpendicular distance (on Figure 5) of the required centre of mass from axis QG_1 is thus

$$\frac{\text{total first moment of mass about } QG_1}{\text{total mass}} = \frac{1400}{1850} = 0.76 \text{ mm.}$$

Since Figure 5 is drawn to a scale of $\frac{1}{48}$ full size, the real distance of the centre of mass of the actual machine from axis QG_1 is

$$48 \times 0.76 = 36 \text{ mm}$$

(there is no need to bother about fractions of a millimetre since all our figures are approximate.)

This leaves the other co-ordinate to be found.

Exercise

Calculate the distance of the centre of mass of the dipper-arm assembly from axis QO (Figure 5).

This time we use the perpendicular distances of the component centres of mass from axis QO, as shown in Table 3.

10

Table 3 First moments of mass for front end equipment components about axis QO

Component	Mass/kg	Perpendicular distance of centre of mass to axis QO/mm (from Figure 5)	First moment of mass about axis QO/kg mm
Dipper arm	300	12	3600
Bucket (full)	1400	52.5	73500
Bucket cylinder	100	16.5	1650
Bucket links	50	37	1850

Total first moment of mass about QO = 80600

Hence the distance from QO of the centre of mass of the assembly (on Figure 5) is

$$\frac{80600}{1850} = 43.6\,\text{mm},$$

and the full-size distance is

$$48 \times 43.6 = 2093\,\text{mm} = 2.09\,\text{m}.$$

We can now plot the position of the centre of mass on Figure 5. This is done in Figure 6(a), which shows the total weight W of the assembly acting through the centre of mass G_T. The line of action of W intersects the line of action of the force at P in point S (the force at P is due to the dipper-arm cylinder).

There are only three static forces acting on the dipper-arm assembly: the third is the force at pin Q. To maintain equilibrium, this force must be equal and opposite to the resultant of the other two forces and must therefore pass through point S. (Since neither the force at P nor the weight can have a moment about S, their resultant cannot have a moment about S). We thus know the directions of all the forces on the dipper-arm assembly and the magnitude of one of them (the total weight), allowing us to draw a force diagram to scale (Figure 6(b)). From this diagram we can measure the magnitude of the total static force at Q: it comes to about 18800 N. The arrowhead in Figure 6(b) refers to the force on the dipper-arm assembly; the corresponding force on the boom of the machine has the opposite arrowhead and is the resultant of the force at P and the weight of the dipper-arm assembly. The force at P arises from the compression of the oil trapped in the right-hand end of the dipper-arm cylinder by the piston which is attached to the piston rod. This, in turn, is attached to the dipper arm at P and is being pulled to the right by the dipper-arm assembly (Figure 6).

SAQ 1

If the distance of point G_T (the combined centre of mass) from point Q is increased by 10%, determine by redrawing the force diagram, the percentage change in the static force on the pin at Q.

So much for the static analysis. We now turn to the effect that acceleration of the dipper-arm ram will have on the force at pin Q. In order to estimate this we leave the assembly in the same position as before and we look at it when the piston of the dipper-arm ram has its greatest acceleration.

2.2.2 Dynamic forces

What is the greatest acceleration of the dipper arm ram?

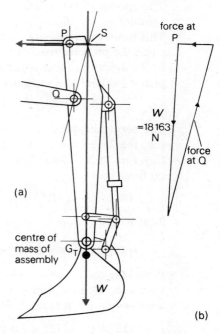

Figure 6   (a) Forces on the dipper-arm assembly: scale 1/48 full size. (b) Vector diagram: scale 1 in ≡ 10^4 N

SAQ 1

To find this we shall assume that the operator of the digger suddenly fully opens the oil valve to the dipper-arm cylinder when the dipper arm is stationary; thus for an instant the whole of the available oil pressure acts on the ram. (Once the dipper arm, and hence the ram, begins to move there is a flow of oil into the cylinder. This will cause the oil pressure in the cylinder to drop because it is necessary to have a pressure drop to push the oil into the cylinder against the friction of the pipes and through the valve; this will be dealt with in more detail later in the course.)

We shall concentrate our attention on the instant immediately after the opening of the valve.

At this instant the force exerted by the oil is transmitted by the ram to the dipper arm through the pin at P. Since this force has a moment about the axis of rotation Q of the dipper arm, the dipper arm (including the bucket etc.) will have an angular acceleration, but it will not yet have had time to acquire a measurable angular velocity. We can therefore treat it as a rigid body, pivoted about a fixed axis (pin Q), which has an angular acceleration but no angular velocity. Alternatively we might say that since the dipper-arm assembly is subject to an extra force at P (that is, in addition to the static forces) and to a corresponding reaction at Q, the centre of mass of the assembly must have an acceleration.

The specification of the machine tells us that the greatest available oil pressure in the dipper-arm cylinder is 13.1 MPa, that is $13\,100\,\mathrm{kN\,m^{-2}}$. This limit is determined by setting the relief valves, which begin to open when the pressure reaches this value. It follows that when the inlet valve to the dipper cylinder is suddenly opened, the instantaneous force on the piston (which has a diameter of 13.3 cm) will be

$$\pi/4 \times (0.133)^2 \times 131 \times 10^5 = 182\,000\,\mathrm{N}.$$

The perpendicular distance from pin Q to the line of action of this force (i.e. the longitudinal axis of the cylinder) is 15.5 mm in Figure 6 and therefore 15.5×48 mm in the machine. It follows that the moment of the oil force is

$$182 \times 10^3 \times 0.0155 \times 48 = 135.4 \times 10^3 = 135\,400\,\mathrm{N\,m}.$$

Similarly the moment about Q of the weight of the dipper-arm assembly is,

$$181\,63 \times 0.003 \times 48 = 2615\,\mathrm{N\,m}.$$

If we take the oil force to have the arrowhead shown in Figure 7, (that is we assume that the ram rod is about to move out of the cylinder) then the total torque about Q on the dipper-arm assembly is clockwise and equal to

$$(135\,400 + 2615) = 138\,015\,\mathrm{N\,m}.$$

Therefore the clockwise angular acceleration of the assembly α (in $\mathrm{rad\,s^{-2}}$) is given by

$$138\,015 = I\alpha \qquad (1)$$

where I is the total second moment of mass about Q of the dipper-arm assembly (in $\mathrm{kg\,m^2}$).

We now want to obtain a reasonable estimate of I. We do this in these stages: first making suitable simplifying assumptions for each of the components; second working out the approximate second moment of mass about Q for each component (based on these assumptions); and third adding up the component values to obtain the total estimated value of I.

For the dipper arm itself we begin by making the same assumption as before: that it has a uniform cross-section throughout its length. The actual and the assumed shapes are shown in Figure 8, the assumed or approximate shape in dotted lines.

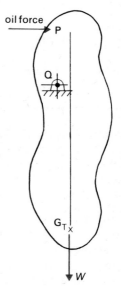

Figure 7

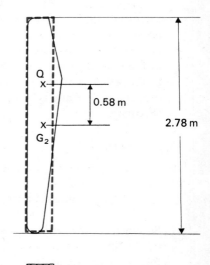

cross-section

Figure 8 The dipper arm: the dotted line shows the assumed shape

The scale drawing suggests another way in which we can simplify the calculation of the second moment of mass. Since the dimensions of the cross-section we are assuming are clearly small compared with the total length of the component, we can treat it as a rod of uniform small cross-section. This is one of the cases listed in Table 1 of Unit 6/7. It shows that the second moment of mass about an axis through the centre perpendicular to the length of the rod is $ML^2/12$, where M is the mass of the bar and L is its length. By using the *parallel axes theorem* we find the second moment about a parallel axis through the axis of rotation of the dipper arm to be

$$M\left(\frac{L^2}{12}+d^2\right)$$

where d is the distance between the two parallel axes. (A proof of the parallel axes theorem is given in the appendix of this unit.)

For the dipper arm $M = 300\,\text{kg}$, $L = 2.78\,\text{m}$ and $d = 0.58\,\text{m}$, so the value for the second moment of mass about a perpendicular axis through Q is

$$300\left[\left(\frac{2.78^2}{12}\right)+(0.58)^2\right] = 300[0.644+0.336]$$

$$= 300 \times 0.980 = 294\,\text{kg}\,\text{m}^2.$$

The details of the corresponding calculations for the other components (and of the simplifying assumptions made in each case) are given in Table 4. The total second moment of mass I is thus $9444\,\text{kg}$ (by adding values in the last column).

Table 4 Calculations for second moment of mass about axis of rotation through Q

Component	Mass M/kg	Simplifying assumptions	Second moment of mass about an axis perpendicular to plane of diagram through the centre of mass /kg m^2	Distance d of centre of mass from Q/m	Second moment of mass about axis of rotation through Q (i.e. perpendicular to plane of diagram) /kg m^2
Dipper arm	300	small uniform cross-section	$300 \times \dfrac{2.78^2}{12} = 193$	$48\sqrt{(0.001^2+0.012^2)}$ $= 0.48\sqrt{(0.01+1.44)}$ $= 0.58$	$193 + [300 \times (0.58)^2]$ $= 294$
Bucket (with load)	1400	mass concentrated at G_1	—	48×0.0525 $= 2.52$	$1400 \times (2.52)^2 = 8890$
Bucket cylinder	100	small uniform cross section	$100 \times \dfrac{(48 \times 0.031)^2}{12}$ $= 18.5$	$48\sqrt{(0.009^2+0.0165^2)}$ $= 0.48\sqrt{(0.81+2.72)}$ $= 0.902$	$18.5 + [100 \times (0.902)^2]$ $= 18.5 + 81.4 = 100$
Bucket links	50	mass concentrated at G_4	—	$48\sqrt{(0.004^2+0.037^2)}$ $= 0.48\sqrt{(0.16+13.7)}$ $= 1.79$	$50 \times (1.79)^2 = 160$

Notice that the bucket is by far the largest contributor to the total: in fact we should not have been far wrong if we had completely neglected all the other components. Thus there is no point in complicating the calculations even further for the sake of a little more accuracy. On the other hand, it is worth looking at the calculation for the bucket again. In the table we treated it as a concentrated mass. In view of the fact that it is fairly compact and a long way from the axis of rotation this seems reasonable (the Md^2

13

term is likely to be much greater than the term relating to the second moment of mass about the perpendicular axis through G_1). You might like to check this for yourself.

SAQ 2

Re-evaluate the second moment of mass of the loaded bucket about the axis of rotation at G_1 on the assumption that the bucket has the shape of a sphere of radius 48 cm with its centre at G_1 (draw this to scale on Figure 1 to convince yourself that the dimension is reasonable).

The solution to SAQ 2 should have convinced you that the value of I obtained from Table 4 is adequate for our purpose and we can now substitute this value in equation (1):

$$138\,015 = 9444\alpha$$

$$\alpha = \frac{138\,015}{9444} = 14.7\,\text{rad s}^{-2}.$$

This is the maximum angular acceleration of the dipper arm that we can expect when the inlet valve to the dipper-arm cylinder is suddenly opened. The calculation of the force at the pin Q at this instant is now straightforward. Figure 9(a) is a repeat of Figure 7 and shows the forces acting on the dipper arm assembly as well as the position of its centre of mass G_T, which is at a distance from Q of

$$\sqrt{(0.036^2 + 2.09^2)} \simeq 2.1\,\text{m}.$$

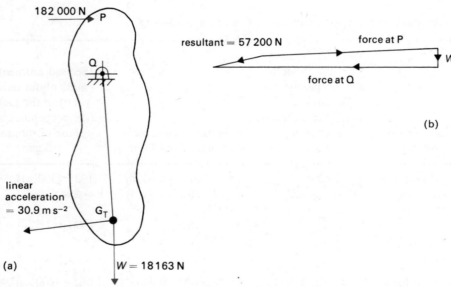

Figure 9 (a) Forces on the dipper-arm assembly. (b) Vector diagram: scale 1 in $\equiv 10^5$ N

Since at the instant we are considering the whole assembly has no angular velocity but only angular acceleration, the total linear acceleration of the centre of mass will equal

$$14.7 \times 2.1 = 30.9\,\text{m s}^{-2}$$

(because 'a = rα') in a direction at right angles to QG_T. Since the total mass of the assembly is 1850 kg, the accelerating force on the assembly (which is the resultant of the external forces) is

$$1850 \times 30.9 = 57\,200\,\text{N}.$$

The diagram of forces on the dipper-arm assembly is shown, to scale, in Figure 9(b). The force at the pin Q, as measured from the diagram, is about 230 000 N, which is about 12 times the static force which we obtained in the last section. Notice the great importance in the force diagram of the force at P due to oil pressure.

Clearly the static force does not give us any useful guidance for the design of the pin at Q (and its bearings in the boom and in the dipper arm), when dynamic forces of this magnitude can be expected to act on it – albeit infrequently and for short periods of time. In any case, repeated shock loads can be a very severe form of loading, producing stresses which are greater than those produced by steady forces of the same magnitude.

We have done this calculation as an example of the possible importance of dynamic forces in machines and to demonstrate that even in a slow-moving machine, where the static forces are large, dynamic forces are not necessarily negligible.

2.3 Dynamic bending moments and stresses

Since the dynamic forces on the dipper-arm assembly have turned out to be important in the case of the sudden loading which we analysed in the last section, it is worth taking a look at the corresponding stresses in the material of the dipper arm. These will be due mainly to bending. In order to estimate them, we will, once again, make some simplifying assumptions.

To begin with, just to see the main idea, imagine that the dipper arm is made of soft rubber. When we give it a hard push at a point above its axis of rotation, it will immediately acquire an angular acceleration and it will bend into the sort of shape sketched in Figure 10(a).

This is also true of the steel dipper arm, but the amount of bending and the strain produced is much less than that with the rubber. (In fact, the distortion is small enough to justify our treatment of the arm as a rigid body).

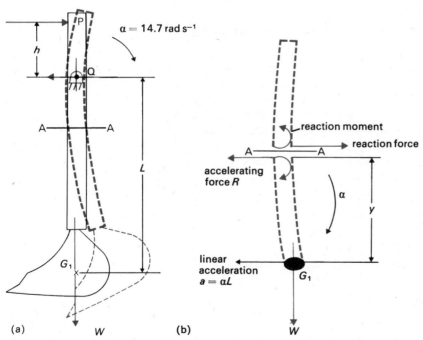

Figure 10   (a) Exaggerated bending of the dipper arm. (b) Effects of acceleration on the dipper arm (the reaction force is the force exerted by the material below on the material above the section AA

We shall continue to treat the loaded bucket as a concentrated mass and Figure 10(b) shows the effect of its acceleration on the rest of the dipper arm. At any section through the dipper arm, the required accelerating force has to be transmitted to the part of the dipper arm below the section considered. This force must be accompanied by a moment, as shown, and the reaction of this moment on the dipper arm above the section is the bending moment which bends it into the shape sketched (in an exaggerated manner) in Figure 10(a). The magnitude of the moment can be deduced by considering the dynamics of the part of the dipper arm below the section considered.

Let us first look at the effect of the bucket, neglecting the mass of the dipper arm itself. If we consider the bucket as a concentrated mass, we ignore its second moment of mass. This means that the total moment of external forces and couples about its centre of mass must be zero. (Remember that torque $= I\alpha = 0$ when $I = 0$, whatever the value of α).

The part of the dipper arm above the section AA exerts a force R on the part of the dipper arm below AA. Since we are neglecting the mass of the dipper arm itself, this force must be just enough to give the bucket its linear acceleration. This acceleration equals αL (see Figure 10) and so

$R = M\alpha L.$

This force, however, is not the only effect produced by the upper part of the dipper arm on the part below AA. Since the total moment about the centre of mass of the bucket must be zero, there must also be a couple at section AA equal to Ry. Its direction and that of its reaction on the part of the dipper arm above AA are shown in Figure 10(b).

A similar argument to that given above also applies to each slice of the dipper arm itself, but here we will continue to neglect the mass of the dipper arm. After all, as we have seen, most of the mass of the dipper-arm assembly is in the bucket. Furthermore, the bucket is at the end of the dipper arm, as far from the axis of rotation as possible, which gives it the largest linear acceleration. For a rough estimate of the dynamic stresses, therefore, the effect of the bucket alone will be good enough, provided we remember that our figures will give an estimate which is somewhat on the low side. Only if the estimated stresses turn out to approach dangerous levels, or even rise to half the level of the failure stresses, will there be any point in complicating the analysis.

What we really want is an estimate of the maximum dynamic bending moment in the dipper arm. We have seen that for values of $y \leqslant L$ this bending moment equals Ry, For values of $y > L$ the effect of the external forces at the bearing will be to cause the bending moment to decrease to zero towards the top end P of the dipper arm. The greatest dynamic bending moment, therefore, is at the section through the bearing Q, when $y = L$. (Have you noticed that the actual width of the dipper arm is greatest at this section? Look at Figure 5 for example.)

We can now put in some figures and get a value for this bending moment. Table 3 shows us that the mass of the bucket $M = 1400\,\text{kg}$ and that $L = 0.0525 \times 48 = 2.52\,\text{m}$. In Section 2.2.2 we worked out that $\alpha = 14.7$ rad s^{-2}. It follows that our estimate of the dynamic force on the bucket is $(1400 \times 14.7 \times 2.52)\,\text{N}$ and the corresponding bending moment at the section through Q is

$[1400 \times 14.7 \times (2.52)^2] = 130\,700\,\text{N m}.$

SAQ 3

SAQ 3

Sketch the diagrams of bending moment and shearing force on the dipper arm due to the accelerating force on the bucket considered, as a point mass. Neglect the mass of the dipper arm and gravity forces.

The full-size dimensions of this section, scaled from the manufacturers' drawings, are approximately as shown in Figure 11. The second moment of area I_A of this section for bending about axis BB (derived in Unit 5) is

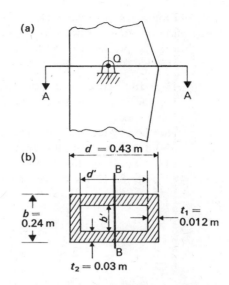

$$\frac{bd^3}{12} - \frac{b'd'^3}{12}$$

where

$$b' = b - 2t_2 = 0.24 - 0.06 = 0.18,$$
$$d' = d - 2t_1 = 0.43 - 0.024 = 0.406.$$

Thus

$$I_A = \frac{1}{12}(0.24 \times 0.43^3 - 0.18 \times 0.406^3)$$

$$= \frac{1}{12}(0.24 \times 0.08 - 0.18 \times 0.067)$$

$$= \frac{0.0071}{12} = 0.00059 \text{ m}^4.$$

Figure 11 (a) Elevation of the dipper-arm about joint Q. (b) View of section AA

The maximum bending stress due to the dynamic bending moment, is given by

$$\sigma = \frac{\text{bending moment}}{I_A} \times \frac{d}{2}.$$

Neglecting the (very small) stress due to the weight of the whole assembly, σ will be the principal stress in the extreme right-hand and left-hand fibres of the section in Figure 11 acting normally to the plane of the diagram.

Substituting the appropriate numbers, we get

$$\sigma = \frac{130\,700}{I_A} \times \frac{0.43}{2} = \frac{130\,700}{0.00059} \times 0.215$$

$$= 10^6 \times 47.6 \text{ N m}^{-2} = 47.6 \text{ MPa}.$$

Since the yield stress for steel is of the order of 200 MPa this value of the maximum bending stress is quite safe, even allowing for the fact that our bending moment is an under estimate. All the same, you might like to check the necessity of enlarging the dipper arm at the section through G by working out the following problem.

SAQ 4

Determine the maximum bending stress corresponding to a bending moment of 130 700 Nm about axis AA at a section which has the dimensions shown in Figure 12. By sketching Mohr's circle, determine the greatest shear stress which corresponds to this maximum bending stress, in magnitude and direction.

SAQ 4

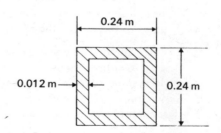

Figure 12

We could go on to analyse other types of movement of the excavator: the effect of acceleration of the boom ram pistons, for example, or the rapid back-and-forth motion which one often sees being given to the buckets of excavators in order to empty them of wet clay soil which tends to stick. We will, however, content ourselves with a brief look at one further dynamic effect that has not yet been mentioned.

2.4 The effect of the tractive force

The prime mover which supplies the power for all the operations of the excavator considered in the case study is a six-cylinder diesel engine. This engine drives hydraulic pumps which supply high-pressure oil to the hydraulic slewing motor and to the cylinders which move the boom, the dipper arm and the bucket, as well as to the hydraulic motors which drive the tracks and which propel and steer the machine as a whole. In deploying the propulsive power, speed has been sacrificed in favour of a high tractive force.

High road speed is clearly unnecessary and undesirable for a vehicle as high and as heavy as this excavator, but the high tractive force is essential for pushing against obstacles, for moving over uneven ground and for climbing up steep slopes. The makers' specification puts the road speed at about $2\,\mathrm{km\,h^{-1}}$ and the tractive force at about $80\,\mathrm{kN}$. Since the total mass of the whole machine is about $12 \times 10^3\,\mathrm{kg}$, this could result in accelerations of up to $80/12\,\mathrm{m\,s^{-2}}$, that is $6.67\,\mathrm{m\,s^{-2}}$, on a horizontal road.

The high tractive force is essential for pushing against obstacles, for moving over uneven ground and climbing steep slopes. . . .

This acceleration is a good deal smaller than the $30.9\,\mathrm{m\,s^{-2}}$ which we had to cope with in the last section. For the position of the mechanism which we looked at there, the acceleration of the combined centre of mass of the dipper-arm assembly was in a nearly horizontal direction. We can thus account for the possibility that the two effects might take place simultaneously by increasing the resultant force in Figure 9(b) by about 20% (assuming the acceleration of the machine as a whole to be from right to left), which would make little difference to the force at the pin Q.

Now consider the effect on pins A and B (Figure 1). When the machine is stationary, the combined effect of these two pins is to support the weight of the boom, the dipper-arm assembly and the hydraulic cylinders. This means that the vector sum of the forces at A and B must be vertical and equal to the weight they support. Let us call this W.

When the machine accelerates along a horizontal road, pins A and B will have to supply the horizontal force necessary to accelerate the mass W/g. (Remember that $W = Mg$ and hence $M = W/g$). The total force due to the two pins will now be W vertically and $(W/g \times 6.67)$ horizontally. The resultant force F_{R} due to the two pins together is therefore

$$\sqrt{\left[W^2 + \left(\frac{W}{g} \times 6.67\right)^2\right]} = W\sqrt{\left[1 + \left(\frac{6.67}{9.81}\right)^2\right]}$$

(Figure 13). It follows that

$$F_{\mathrm{R}} = W\sqrt{(1+0.46)} = 1.21W$$

which is about 20% more than the static force. This is worth bearing in mind. So is the possibility of vertical accelerations due to uneven ground: these are very difficult to predict but will in our case be conveniently limited by the very low road speed which we have already mentioned.

Last, we must look at the operation of climbing slopes. What is the steepest gradient that the machine will be able to climb?

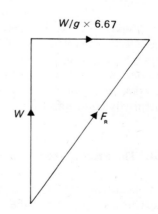

Figure 13   Resultant of forces due to pins A and B

Figure 14 represents the machine moving up an incline which makes an angle θ with the horizontal. F is the tractive force and W is the weight of the whole machine. The equation of motion along the incline is

$$F - W \sin \theta = (W/g)a,$$

where a is the acceleration of the machine up the incline. As θ is increased, $\sin \theta$ increases (θ is clearly much less than 90°); therefore the left-hand side decreases and with it the value of a on the right-hand side. The greatest slope the machine can manage is that at which $a = 0$, where the machine can just maintain motion up the slope with no acceleration. For this case (assuming that the machine does not topple over),

$$F - W \sin \theta = 0,$$
$$F = W \sin \theta,$$
$$\sin \theta = F/W.$$

For the figures quoted at the beginning of this section, this means that

$$\sin \theta = \frac{80 \times 10^3}{12 \times 10^3 \times 9.81} = \frac{80}{118} = 0.68.$$

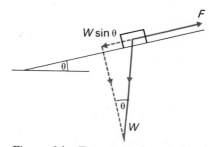

Figure 14 Forces on the machine when it is climbing

Hence θ is about 42°. Of course, if the machine came to a halt on such a slope, it could not start again (except to move downwards) because there is no tractive force left for upward acceleration. We should therefore be wise to allow a safety margin, and indeed the specification of the machine states that the maximum gradient is 1 in 1.6. (This is equivalent to saying that $\sin \theta = 1/1.6 = 0.625$, that is $\theta = 38.7°$.) This is a very steep gradient for a road vehicle to climb (a gradient of 1 in 8 is considered steep for a road which carries normal motor traffic) and the ability to climb such slopes is an important feature of the excavator.

SAQ 5

Calculate the acceleration of the excavator and the power output of the engine at an instant when it moves up a gradient of 1 in 5 at a speed of $1.92 \, \text{km h}^{-1}$, assuming that it exerts its greatest tractive force (80 kN).

SAQ 5

2.5 A look back and a look ahead

We began our treatment of dynamic forces by analysing some of the movements of the excavator from the case study. On the whole, the dynamic forces in this machine turned out to be small in value compared with the static forces (with the exception of those discussed in Sections 2.2.2 and 2.3) and those associated with normal lifting were found to be negligible. This is because the machine consists of massive, slow-moving links: we could reasonably expect quite a different result from the analysis of a fast-moving machine with light components subject to considerable accelerations.

Let us look at such a machine. We do not need to depart very far from the excavator for this, because the diesel engine that provides its power is a good example of a relatively fast-moving machine in which the gravity forces are unimportant compared with the dynamic ones. The components of the engine are of many different types and will provide practical examples for revising and applying more of the material from earlier parts of the course.

19

Not Examinable.

THE RECIPROCATING INTERNAL COMBUSTION ENGINE

This type of engine converts some of the internal energy in hydrocarbon fuel into mechanical energy and useful work in the following way: it burns the fuel inside each cylinder (hence the name 'internal combustion') and then uses the slider–crank mechanism which you met in Unit 3 to convert the straight-line oscillatory motion of the piston (hence 'reciprocating engine') into rotary motion of the output or crank shaft. The details of the processes that take place in each cylinder and the energy relationships between the fuel and the output torque and speed will be dealt with in later units. For the present, let us concentrate on the basic mechanism and on the forces and torques which it transmits: that is, on the basic *machine*.

To study this, we shall begin by looking at *one* of the cylinders of the engine. Each cylinder has its own slider–crank mechanism; all of them are exactly the same except that their cycles of operations are out of phase with one another. For the engine we are considering, each cycle of operations takes place during two complete rotations of the crankshaft or four strokes of the piston (hence it is called a *four-stroke engine*).

inlet valve open   spark   exhaust valve open

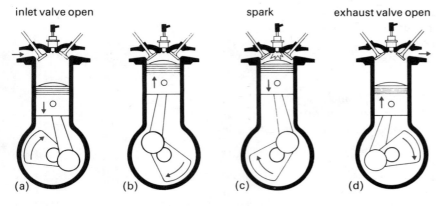

(a)   (b)   (c)   (d)

Figure 15 Operation of a four-stroke internal combustion reciprocating engine. (a) Suction stroke: petrol/air mixture is drawn into the cylinder. (b) Compression stroke: mixture is compressed by the rising piston. (c) Expansion stroke: burning mixture drives the piston down. (d) Exhaust stroke: rising piston pushes out the burnt gases

For a petrol or spark ignition engine, these operations are shown diagramatically in Figure 15. This figure also illustrates the action of the inlet and outlet valves, which are operated by cams so as to open and close at the same point in every cycle. We shall have more to say about the design of such cams later in this unit. A four-stroke diesel or compression ignition engine works on the same general principles except that it uses a different fuel. This fuel is pushed (rather than, as in petrol or spark-ignition engines, sucked) into the cylinder to mix there with the air which has previously been drawn in by the piston, and compression effects cause the mixture to burn without the use of an electrical sparking plug. Although the present discussions will centre only on four-stroke engines, note that not all reciprocating engines operate in this way.

Let us consider the forces which arise in the operation of one cylinder of a four-stroke engine. To keep the discussion specific and realistic we shall look at one of the six cylinders of the diesel engine in our excavator.

3.1 Gas force and turning moment

Gas force is the name that we shall give to the force on the piston due to the gases in the cylinder at any moment. The word 'gases' will be understood to include the air–fuel mixture which will occupy the cylinder during the compression stroke. We now look at the relationship between the gas force and the torque on the crankshaft at any moment during the operation of the engine.

Figure 16 shows the slider–crank mechanism at an instant when the gas force in the cylinder is F_G. To begin with, we shall neglect the masses of the components of the engine. We therefore treat the connecting-rod as a light structural link loaded only at its ends and this means that if there is little friction at its end bearings, the loading on the connecting-rod will be purely axial. Since it has to make the crank rotate against the external load torque C, it is clearly in compression. We will call the compressive force on the connecting-rod Q.

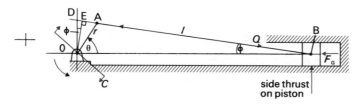

Figure 16   Forces on the piston and crank

The total moment of the forces on the crank about the crankshaft centre O is equal to

$$Q \times (OE) - C = 0$$

(for light links resultant forces and moments must be zero), and since OE = OD cos ϕ, we can write

$$Q \times (OD) \cos \phi = C. \tag{2}$$

At the piston end, the effect of the force Q is two-fold: its component along the line of stroke balances the gas force, that is

$$Q \cos \phi = F_G; \tag{3}$$

and its components at right angles to the line of stroke produces a *side thrust* between the piston and the cylinder. This side thrust will equal $Q \sin \phi$ and causes friction and wear between the piston and the cylinder. Substituting for $Q \cos \phi$ from equation (3) into equation (2) we get

$$F_G \times (OD) = C = \text{external turning moment at the crankshaft.}$$

We now have an expression for the turning moment due to the gas force, neglecting the effect of the masses of the engine components.

The most important feature of this relationship lies in the fact that both the quantities on the left-hand side vary continuously as the engine runs. The gas force varies from a very high positive value at the moment of explosion to a slight negative value during suction (negative because the pressure in the cylinder will then be below atmospheric and the effect will be to retard the motion of the piston). The distance OD is the intercept of the centre line of the con-rod (produced if necessary) on the crank centre line perpendicular to the line of stroke. It varies, approximately, between the values of $+R$ and $-R$. (When D is below O, that is when the piston is moving to the right in Figure 16, the intercept of OD is taken as negative. Thus during the compression stroke, for example, the available turning moment is negative and each cylinder requires a power *input* during this part of the cycle. The same argument holds for the exhaust stroke, though to a lesser extent.)

In short, since both F_G and OD vary widely during each cycle, it follows that the turning moment produced by each cylinder of a reciprocating engine fluctuates sharply during each cycle. This is one of the great drawbacks of the reciprocating engine which has to convert straight line motion into rotary motion. A fluctuating output torque means that at some parts in the cycle the engine torque is greater than the torque required by the load, and at other points it is smaller. This causes angular accelerations and decelerations, thereby producing dynamic stresses, and makes it impossible to maintain a truly constant speed.

In a multi-cylinder engine the torque variations in the different cylinders cancel out to some extent and the resulting total engine torque is more uniform than that due to any one cylinder, but is still far from constant.

SAQ 6

For each cylinder of the diesel engine in the excavator, $r = 0.064$ m, $l = 0.22$ m and the diameter of the piston is 0.098 m. Consider one cylinder only and assume that when $\theta = 30°$ the gas pressure in the cylinder is 200×10^4 N m^{-2}. Neglecting the masses of the engine components and the pressure on the outside of the piston, estimate the turning moment due to that cylinder at that instant. (The easiest way to find the required intercept is to draw the mechanism to scale.) If the angular velocity of the crankshaft at the relevant instant is 2400 rev min^{-1}, what is the instantaneous power output of the engine?

SAQ 7

Repeat the calculation in SAQ 6 for $\theta = 150°$, assuming that for this position

$$F_G = 2 \times 10^3 \text{ N}.$$

3.2 Engine dynamics

We now go on to analyse the forces which were omitted in the last section: those that are required to accelerate the moving parts of the engine. A complete force analysis will then be obtainable by superimposing these forces and those discussed in the last section.

We shall neglect gravity forces as being too small to matter. The connecting-rod of the diesel engine in the excavator of the case study, for example, weighs only about 23 N; the forces we shall deal with will all be several orders of magnitude greater than this. (To explain the meaning of *order of magnitude* we can compare the numbers 2×10^3 and 4×10^3: these are of the same order of magnitude, while 5×10^5 is two orders of magnitude greater.)

There are three sets of moving parts in the engine: the piston (with gudgeon pin and piston rings), the connecting-rod (with its bearings), and the crank (with its crank pin). We shall look at the motion of each of these and the forces associated with the motion.

3.2.1 The piston assembly

The piston moves back and forth along a straight line (known as the *line of stroke*) with the acceleration which was derived in the appendix to Unit 3. The approximate form of this acceleration is

$$a = \omega^2 r \left(\cos \theta + \frac{r}{l} \cos 2\theta \right). \tag{4}$$

The fairly high degree of accuracy of this expression was assessed in Unit 3. It assumes that the angular velocity ω of the crank is uniform:

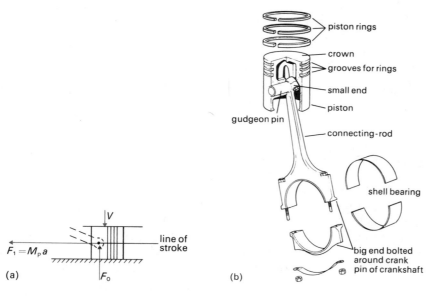

piston rings
crown
grooves for rings
small end
piston
gudgeon pin
connecting-rod
shell bearing
big end bolted
around crank
pin of crankshaft

$F_1 = M_p a$
line of stroke
V
F_0

(a)

(b)

Figure 17 (a) Forces on the gudgeon pin. (b) Piston and connecting-rod assembly

this is never quite true in practice but it is quite possible to keep the variation of ω within, say, 1% or less of its mean value by the use of a flywheel, as we shall see later.

The accelerating force required for the piston assembly is therefore equal to $M_p a$ along the line of stroke, where M_p is the total mass of the piston assembly, and this force must come from the action of the small end of the connecting-rod on the gudgeon pin, as shown in Figure 17(a). The other component F_0 of the force due to the connecting-rod, is opposed by the lateral reaction V of the cylinder wall. Figure 17(b) is a picture of the piston assembly and shows the way in which it is joined to the connecting-rod and its constituent parts.

SAQ 8

SAQ 8

Using the data in SAQ 6, calculate for the instant when $\theta = 30°$:

(a) the angular velocity ω rad s^{-1} of the crank;

(b) the acceleration of the piston;

(c) the accelerating force F_1 on the piston due to the connecting-rod, the total mass of the piston being 2 kg.

3.2.2 The connecting-rod

As we saw in Section 3.2.1, the centre of the *small end* of the connecting-rod moves along the line of stroke with the acceleration a given by equation (4). The *big end* consists of a bearing which fits around the crank pin so that its centre moves in a circular path about the axis of the crank-shaft. On the basis of our assumption that the angular velocity ω of the crank is practically constant, the acceleration of the centre of the big end is $\omega^2 r$ towards the axis of rotation.

These accelerations are shown in Figure 18(a) together with that of the centre of mass G of the connecting-rod, which is intermediate in direction between the accelerations of the big-end and small-end centres. Figure 18(b) is a sketch of the acceleration diagram for the engine mechanism, from which it can be seen that, at the instant shown, the angular acceleration α of the connecting-rod is anti-clockwise. (Check this yourself, looking back at Unit 3 if necessary. At the same time, check that the magnitude of α is b″b′/AB.) The vector o′b′ represents the piston acceleration, o′g′ is the acceleration of G, and o′a′ is the acceleration of the big-end centre.

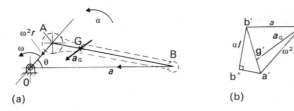

(a)   (b)

Figure 18 Acceleration diagrams for the big end, small end and crank

The resultant force on the connecting-rod equals $M_c a_G$, where M_c is the mass of the rod. This force is parallel to a_G and has the same arrowhead, but its line of action is displaced from G by a perpendicular distance equal to $K_G^2 \alpha / a_G$ where K_G is the radius of gyration of the rod about a perpendicular axis through G. (The argument leading to this statement is given in detail in Units 6/7). It follows that the forces on the connecting-rod will be as shown in Figure 19, where F_0 and F_1 are the dynamic forces on the con-rod exerted by the piston (equal and opposite to the forces *on* the piston, by Newton's third law) and F_2 is the force on the con-rod due to the crank.

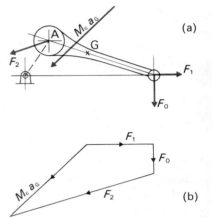

Figure 19 Forces on connecting-rod

At this stage of the analysis we know F_1 and we can determine a_G by drawing the acceleration diagram for the connecting-rod. (This will also act as a check for the piston acceleration evaluated from equation (4)). Hence we obtain $M_c a_G$, leaving only F_0 and F_2 to be found.

The easiest way to find these is to take moments of the forces $M_c a_G$ and F_1 about the big-end centre A (this cuts out F_2). The sum of the moments of F_0 and F_1 must then equal the moment of $M_c a_G$ (since $M_c a_G$ is the resultant force and will have the same moment about any point as the constituent forces). This gives us the moment of F_0 and since we know the line of action of this force, we can find its magnitude. We can now draw the force diagram, Figure 19(b), and this will completely define F_2.

Exercise

Show that in Figure 19, the resultant of the dynamic forces F_0 and F_1 does not act along the line of centres of the connecting-rod.

Let F be the vector resultant of F_0 and F_1. Then, in Figure 19, the force $M_c a_G$ is the resultant of F and F_2 and its line of action must pass through the point of intersection of the lines of action of F and F_2. (Since the moment about this point of each of the forces F and F_2 is zero, that of their resultant must also be zero.)

In other words, the line of action of the force F will pass through the point of intersection of the forces F_2 and $M_c a_G$, not along the line of centres of the connecting-rod.

F_2 directly affects the turning moment produced by the engine, as well as stresses in the crank and con-rod. In practice it will therefore be necessary to evaluate F_2 for many different positions of the engine mechanism in order to plot graphs of turning moment or to find maximum values of stress. Using the method outlined in this section, the process would be rather laborious, mainly because a new acceleration diagram would have to be plotted for each position. This is why it is usual to use simpler, if less exact, methods. One of these is described in Section 3.2.4.

3.2.3 The crank

Figure 20 shows the force F_2 on the *crank* exerted by the connecting-rod. It is equal and opposite to the force F_2 on the con-rod shown in Figure 19. In this position of the engine mechanism, therefore, the effect of having to accelerate the moving parts, that is the piston and con-rod, is to produce a force on the crank which opposes its motion.

There will, however, be other positions of the crank for which the corresponding torque will be the other way and will help to drive the crank.

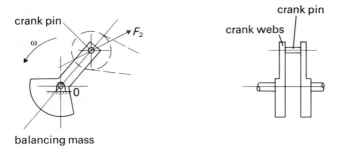

Figure 20   Forces exerted on the crank by the connecting-rod

balancing mass

Notice the *balancing mass*; its purpose is to compensate for the eccentricity of the crank webs and the crank pins and it is so dimensioned that the centre of mass of the whole crank coincides with its axis of rotation. This centre of mass is therefore stationary so that the resultant *force* on the crank is always zero.

The total turning moment on the crank can now be obtained by super-imposing (that is adding) the *static* moment due to the gas force (derived in Section 3.1) and the *dynamic* moment due to F_2. Before we actually do this, however, let us look at the simplified method of analysing the accelerating forces for the con-rod which we mentioned in the last section. This simplified method is widely used in engine design and makes the calculation of the forces in the engine and of the turning moment much easier. The advantage is obtained by making a small approximation.

3.2.4   An easier (though approximate) way to analyse the forces on the connecting-rod

Suppose we replace, conceptually, the mass of the rod by two imaginary concentrated masses joined by a rigid massless link. The most convenient way to do this is to place the two masses at the big-end centre and small-end centre respectively, because the accelerations at these points are easy to calculate and therefore the corresponding forces will be easy to obtain.

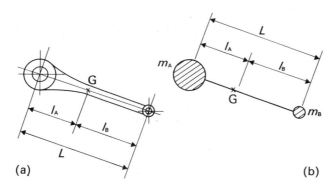

(a)   (b)

Figure 21   (a) Connecting-rod of mass M_c. (b) Equivalent system

We make the sum of the concentrated masses equal to the total mass of the connecting-rod. In addition, we can arrange the masses so that their centre of mass is at the same place as the centre of mass of the connecting-rod. This will mean that the resultant force on our simplified system will be exactly equal to the resultant force on the actual connecting-rod. Let the two concentrated masses in the equivalent system be m_A and m_B (Figure 21). If the equivalent system is to have the same mass as the connecting-rod, then

$$m_A + m_B = M_c. \tag{5}$$

If the centre of mass is to remain in the same position relative to the big-end and small-end centres, then

$$m_A l_A - m_B l_B = 0$$

that is

$$\frac{m_A}{m_B} = \frac{l_B}{l_A}. \tag{6}$$

If equations (5) and (6) apply, the resultant force required for a given acceleration of the centre of mass is the same in magnitude and direction for the equivalent system as for the real connecting-rod. Where, then, does the approximation come in? We find out this when we look at the rotary motion of the connecting-rod.

By fixing the magnitudes and the positions of m_A and m_B we have automatically fixed the second moment of mass of the equivalent system. In fact it will equal $(m_A l_A^2 + m_B l_B^2)$ about an axis through G perpendicular to L, and there is no reason why this should be equal to the corresponding second moment of mass of the actual connecting-rod (except by a coincidence). If it is not equal, then the line of action of the resultant force for the equivalent system will not coincide with that for the actual system (even though the magnitudes and directions of these forces will be exactly the same), since the values of K_G^2 and hence of $K_G^2 \alpha / a_G$ will be different (see Section 3.2.2).

It follows that if a_G and α for the two systems are the same, the corresponding forces in the two systems are not identical: the two systems are not *dynamically equivalent*. However, it is found in practice that the discrepancy is small enough to be tolerated for the sake of the considerable simplification which results from being able to use the following approach.

Add m_A to the mass of the crank pin; increase the balancing mass to account for this, and to bring the centre of mass of the rotating parts back to the axis of the crankshaft; also add m_B to the mass of the piston assembly. The part of the connecting-rod between these masses is taken to be massless and rigid. The extent to which this makes the calculations easier can be seen by referring back to Section 3.1 and Figure 16.

We can immediately write

$$F_G - Q \cos \phi = (M_p + m_B)a$$

so that

$$Q \cos \phi = F_G - (M_p + m_B)a \tag{7}$$

and the torque at the crankshaft is given by

$$C = Q(\text{OD}) \cos \phi$$
$$= [F_G - (M_p + m_B)a] \times (\text{OD}) \tag{8}$$

The term $(M_p + m_B)a$ represents the effect of the masses of the moving parts of the engine and will be negative when a is positive (as shown in Figure 16) and vice versa. As we saw before, during some parts of each cycle the effect of these masses is to reduce the turning moment, and during other parts their effect is to increase it. The quantity $(M_p + m_B)$ is called the *mass of the reciprocating parts* and we shall write

$$(M_p + m_B) = M_{rec}.$$

Figure 22 shows a typical graph of C against crank angle as given by equation (8) for a single cylinder four-stroke engine.

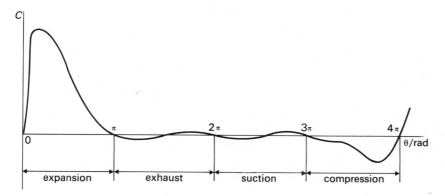

Figure 22   Graph of C against crank angle for a single cylinder four-stroke engine

Exercise

The following data apply to each connecting-rod of the engine which powers the excavator of our case study:

$$L = 0.22\,\text{m}, \qquad l_B = 0.15\,\text{m}, \qquad M_c = 2.3\,\text{kg},$$

where the notation is as in Figure 21.

The mass of each piston assembly is $M_p = 2\,\text{kg}$. Calculate M_{rec}, the mass of the reciprocating parts for each cylinder.

We know that

$$m_A + m_B = 2.3\,\text{kg},$$

and that

$$l_A + l_B = 0.22\,\text{m}$$

therefore

$$l_A = 0.22 - 0.15 = 0.07\,\text{m}.$$

Also, from equation (6),

$$\frac{m_A}{m_B} = \frac{l_B}{l_A}$$

$$= \frac{0.15}{0.07} = 2.14.$$

Thus

$$m_A + m_B = (2.14 + 1)m_B = 2.3,$$

$$m_B = \frac{2.3}{3.14} = 0.733\,\text{kg},$$

and

$$M_{rec} = M_p + m_B = 2 + 0.733 = 2.733\,\text{kg}.$$

Subject to the approximation already mentioned, the mass effect of the con-rod and piston assembly can now be accounted for by imagining that mass m_A is concentrated at the big-end centre and mass M_{rec} at the small-end centre of the con-rod, and by treating the part of the con-rod between these points as a massless rigid bar.

SAQ 9

For the data in SAQ 6, using the results of SAQ 6, SAQ 8, and the exercise above, calculate the turning moment due to one cylinder for which $\theta = 30°$. Take into account the masses of the moving parts.

In a similar manner, we can use our approximate analysis to calculate the forces on the crank. Figure 23 shows the crank complete with balancing mass.

Suppose that the centre of mass of the crank webs, crank pin and the portion m_A of the connecting-rod mass is at G_1, and that the centre of mass of the balancing mass is at G_2. Let the corresponding masses be M_1 and M_2 respectively. Then if the combined centre of mass is to coincide with the axis of rotation through O, we must have

$$M_1 r_1 - M_2 r_2 = 0.$$

Assuming this to be the case, the resultant force on the assembly is zero (since O is a fixed point) and so the force Q at A must be balanced by an equal and opposite force at O, due to the bearing which forms part of the frame of the engine. The couple produced by these two forces is balanced by that exerted by the load (that is, the resisting torque due to the load that the engine is driving). This is shown as C in Figure 23 and will have the value given by equation (8). All this assumes, as before, that the angular velocity of the crank is constant.

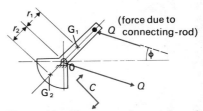

Figure 23   Forces on the crank and balancing mass

Exercise

Figure 24(a) shows part of the crank of a single-cylinder engine. Assuming that the force Q on the crank pin is along the line of centres of the con-rod, identify the stresses at sections AA and BB when the crank angle θ (between the crank and the line of stroke) is (i) $0°$ and (ii) greater than $0°$ but less than $90°$.

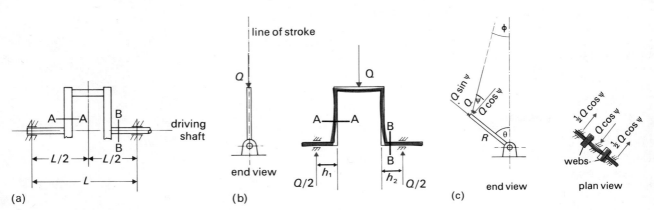

Figure 24   (a) Part of the crankshaft for a single-cylinder engine. (b) Forces on crankshaft, $\theta = 0$. (c) Forces on the crankshaft $0 < \theta < 90°$. (The angle which Q makes with the crank webs is ψ, the Greek letter psi.) The red shapes indicate the deformation under load

(i) $\theta = 0°$ (Figure 24(b)).

At section AA in the crank webs.

There is a compressive stress normal to AA, there are bending stresses normal to AA (bending moment $\frac{1}{2}Qh_1$) and there are shear stresses in the thickness of the web.

At section BB in the crankshaft.

There are bending stresses normal to BB (bending moment $\frac{1}{2}Qh_2$) and shear stresses in the plane of section BB.

(ii) $0 < \theta < 90°$ (Figure 24(c)).

The force Q may be resolved into two components: $Q\cos\psi$ and $Q\sin\psi$, where $\psi = (90° - \theta - \phi)$, if all angles are expressed in degrees rather than radians. Stresses due to $Q\sin\psi$ (along the crank webs) will be similar to those at section AA when $\theta = 0°$, as given above. Stresses due to $Q\cos\psi$ (at right angles to the crank webs) are as follows.

At section AA in the crank webs.

There are bending stresses normal to AA (in the same direction as before, although their maximum occurs in surfaces perpendicular to those of the previous case). There are shear stresses in the plane of section AA: these are partly due to the transverse force and partly due to torsion of each crank

28

web in the plane of AA (the torsion is caused by the twisting moment of the force $\frac{1}{2}Q \cos \psi$ as shown in the plan view of Figure 24(c)).

At section BB in the crankshaft.

There are bending stresses normal to section BB, and shear stresses due to torsion and due to lateral load in the plane of section BB (torque = $RQ \cos \psi$).

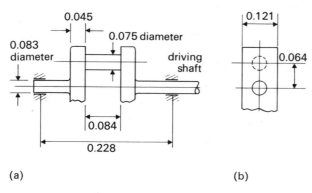

Figure 25   Part of the crankshaft, showing dimensions

Figure 25 shows part of the crankshaft of a single-cylinder engine; all the dimensions given are in metres. The crankshaft may be assumed to be symmetrically supported in the two bearings as indicated.

Assuming that when $\theta = 0°$, $Q = 40$ kN, and treating this as a concentrated force at the centre of the crank pin, estimate the magnitude of the greatest bending stress in the crank pin and state where it occurs. (The relevant second moment of area of a solid circular section of diameter d is $\pi d^4/64$.)

Faulty machining of lubrication ducts in the crank pin gives rise to an estimated bending stress concentration factor of 3. If the greatest permissible shear stress is 100 MPa, can the existing crank pin still be used?

For the engine whose crankshaft is shown in Figure 25, the ratio

$$\left[\frac{\text{crank radius}}{\text{con-rod length}} \right] = 0.29.$$

It is found that the greatest turning moment is 1200 N m and this occurs when $\theta = 30°$. Calculate:

(a) the component perpendicular to the crank radius of the force at the crank pin;

(b) the component along the crank radius of the force at the crank pin;

(c) the greatest shear stress due to torsion in the driving shaft (the relevant second moment of area for a solid circular section of diameter d is $\pi d^4/32$).

3.3   Forces on the engine frame

From the work we have done so far, we can deduce the forces that act on the engine frame due to a combination of the gas force and the dynamic forces of the reciprocating parts in one cylinder of a reciprocating engine.

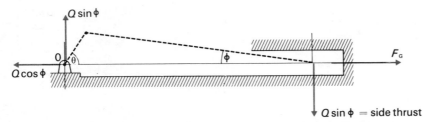

$Q \sin \phi$

$Q \cos \phi$

F_G

$Q \sin \phi =$ side thrust

Figure 26   Forces on the frame exerted by the moving parts of the machine at the piston end and at the crank bearing

We draw the frame by itself and put in the net forces exerted upon it by the moving parts of the machine at the piston end and at the crank bearing (using Newton's third law where appropriate). This gives us Figure 26. The force Q at O has been resolved into components along and perpendicular to the line of stroke, and gravity forces have, as before, been omitted.

The two equal and opposite forces at right angles to the line of stroke constitute the reaction on the engine frame of the load torque C.

The two forces along the line of stroke, however, are not equal, the net force acting to the right in Figure 26 being

$$F_G - Q \cos \phi$$

which from equation (7) is equal to

$$F_G - F_G + M_{rec}a = M_{rec}a.$$

From equation (4), this gives the net force acting to the right as

$$M_{rec}\omega^2 r\left(\cos\theta + \frac{r}{l}\cos 2\theta\right) \tag{9}$$

Equation (9) indicates that the frame is subject to a force along the line of stroke equal to the accelerating force of the reciprocating parts. This force is made up of two components: the *primary* force

$$M_{rec}\omega^2 r \cos\theta$$

and the *secondary* force

$$M_{rec}\omega^2 \times \frac{r^2}{l}\cos 2\theta.$$

Both are periodic, the secondary force having a smaller amplitude than the primary (there is an extra factor of r/l) and twice the frequency.

The effect of these forces is to shake or stretch and compress the engine frame backwards and forwards, and the resulting vibrations are the source of alternating stresses which can lead to fracture, to metal fatigue and to noise and passenger discomfort when the engine is used in a vehicle. This is another of the inherent drawbacks of a reciprocating engine.

The problem of reducing or getting rid of these forces is a well known one. Not much can be done in practice in the case of a single-cylinder engine. In multi-cylinder engines, however, it is possible to arrange the cylinders so that the forces due to the different cylinders cancel out to some extent, and sometimes completely.

Not Examinable.

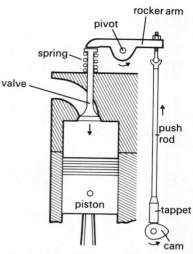

Figure 27   IC engine valve mechanism

MORE ABOUT CAMS

The subject of this section, like the last, is introduced by considering the reciprocating engine. However, this is only one of its many natural applications. We first met and studied cam mechanisms in Unit 3. We also made a brief reference to them in Section 2.2 of this unit, in connection with the four-stroke engine; Figures 15 and 27 illustrate two variations of this particular use of a cam.

One of the simplest cams is illustrated in Figure 28. In this case the cam (1) makes a movement, and the point follower (2) makes a corresponding movement depending on the slope of the cam surface. There is a simple relationship between the movements of cam and follower:

$$y = x \tan \theta.$$

If the operation is to be repeated, a second identical wedge (3) could be placed after 1, as shown. However, it would be more convenient to bend the cam around the z axis (perpendicular to the paper at point O), to achieve the same result (Figure 29).

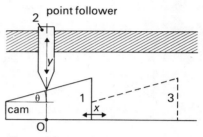

Figure 28

dwell

We can now convert the rotational motion of link 1 into a reciprocating motion of link 2.

Figure 30 shows the meaning of some of the most important technical terms used in cam design. Note that during a *dwell* period, the follower remains stationary. Usually the cam and follower are in line contact only, and so constitute a higher pair. (In practice this means that the contact area is small, because cam and follower are bound to deform, however little). This may have practical implications which cause difficulties, as contact pressures and lubrication must be carefully regulated in order to reduce the wear of the cam surface.

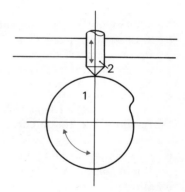

Figure 29 A simple cam with point follower

Most cams rotate at constant angular velocity so that equal angular displacements take place in equal time intervals. But the follower must

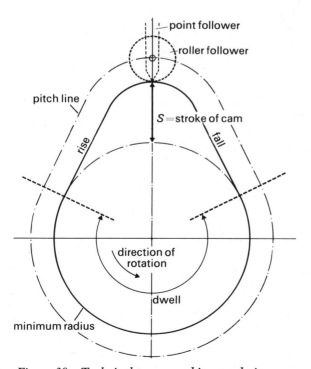

Figure 30   Technical terms used in cam design

return to its starting position at least once during each cycle. Hence its velocity must vary, which means that there must be accelerations. Inevitably the follower and other parts of the mechanism have mass. Consequently forces are generated and the machine must be designed to accept these forces.

In the rest of this section, we look at some of the available methods which enable us to predict, and keep within acceptable limits, the velocities, accelerations and forces corresponding to given displacements.

4.1 Displacement, velocity, and acceleration–time curves

Suppose we require a mechanism to open a poppet valve by a certain amount for a certain time, to close it, and then to repeat the sequence of events. This is what happens in an internal combustion (IC) engine.

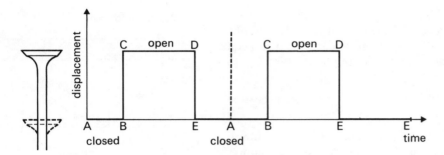

Figure 31 Displacement–time graph for the ideal poppet valve operation

Ideally (Figure 31), we should like the valve to be closed fully from A to B, open fully from C to D, and closed again from E to A until the commencement of the next cycle. This would mean an instantaneous movement from B to C and from D to E.

However, this is just not possible. Not even in a crude engineering model can we assume that a particle of matter can move with infinite velocity. We could settle for a constant rate of displacement, that is constant velocity, by allowing the valve to open uniformly over a period of time (Figure 32), but we should still encounter infinite accelerations of the valve in jumping instantly from one velocity to another, and this of course means infinite forces.

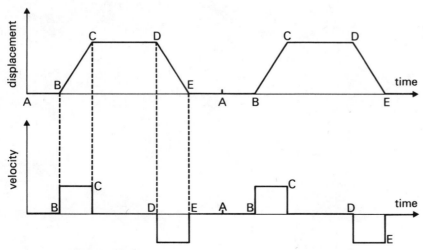

Figure 32 Displacement–time and velocity–time graphs for constant velocity in a poppet valve operation

32

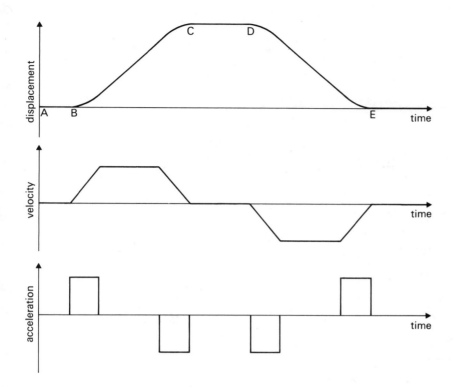

Figure 33 Displacement–time, velocity–time and acceleration–time graphs for a practical poppet valve operation

Thus let us fix on the force level, which we can choose, and look at the acceleration, velocity and displacement diagrams. Constant force applied to the valve or follower means constant acceleration, giving the bottom graph in Figure 33.

We thus arrive at the solution represented in the figure, which is a compromise on that which we originally intended but is at least workable. The forces and accelerations now have finite values, but at the price of the valve taking more time to open and only being fully open for about 50% of the time originally desired.

Let us examine the displacement curve BCD in Figure 34.

We can consider this part of the cycle in three distinct stages:

1 the acceleration of the cam follower away from the cam centre;

2 the motion of the cam follower away from the cam centre at constant velocity;

3 deceleration of the follower as it approaches the limit of its travel.

Period 1, $t_1 \leqslant t \leqslant t_2$:

acceleration $= a_a$ (constant);

$v = a_a(t - t_1)$ thus $v_{max} = a_a(t_2 - t_1)$;

$s = \frac{1}{2}a_a(t - t_1)^2$ which is the equation of a parabola,

$s_a = \frac{1}{2}a_a(t_2 - t_1)^2$.

Period 2, $t_2 \leqslant t \leqslant t_3$:

acceleration $= 0$;

$v = v_{max} = a_a(t_2 - t_1)$;

$s = v_{max}(t - t_2)$ which is the equation of a straight line,

$s_c = v_{max}(t_3 - t_2)$.

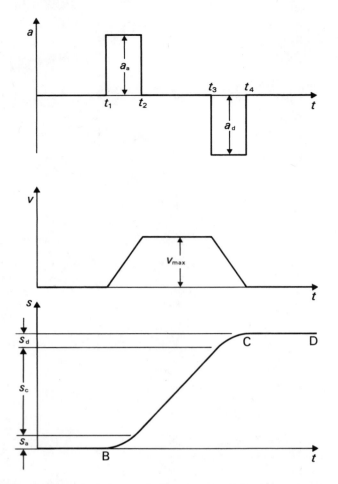

Figure 34   Acceleration–time, velocity–time, and displacement–time graphs for section BCD *of Figure 32*

Period 3, $t_3 \leqslant t \leqslant t_4$:

$$\text{acceleration} = -a_d \quad \text{(constant)};$$

$$v = v_{max} - a_d(t - t_3),$$

$$v_{min} = 0 = v_{max} - a_d(t_4 - t_3), \quad \text{thus} \quad v_{max} = a_d(t_4 - t_3);$$

$$s = v_{max}(t - t_3) - \tfrac{1}{2}a_d(t - t_3)^2, \quad \text{which is the equation of a parabola,}$$

$$s_d = v_{max}(t_4 - t_3) - \tfrac{1}{2}a_d(t_4 - t_3)^2$$

$$= v_{max}(t_4 - t_3) - \tfrac{1}{2}v_{max}(t_4 - t_3)$$

$$= \tfrac{1}{2}v_{max}(t_4 - t_3),$$

which is obvious from the triangular area of this part of the v–t graph.

Thus over a period $t_4 - t_1$ the follower has moved a distance

$$S = s_a + s_c + s_d.$$

On the credit side we now have a cam follower which has:

(a) finite accelerations and hence controllable forces;

(b) a predictable velocity;

(c) a displacement that is composed of three simple sections: two parabolic sections and one straight line section.

However, another look at the acceleration–time curve shows that the resultant forces are abruptly applied and abruptly removed. This means suddenly applied loads which will cause shocks, noise, and vibrations, with the cam and follower behaving like a complex spring–mass system.

These adverse effects, particularly in applications involving medium and high speeds (say over 500 rev min^{-1}), are the basic cause of many failures of machines with cam mechanisms. Sometimes the resulting high contact stresses cause a breakdown of the lubricating film of oil, simply pushing the lubricant away from the surface, allowing metal to metal contact and causing catastrophic wear. Another effect is metal fatigue resulting from rapid reversal of stress on a loaded member.

In fact, the parabolic cam profile and its variants are used mainly in low speed applications. However, better cam forms may be devised.

Consider for instance the trapezoidal acceleration curve shown in Figure 35. This is an improvement, first because there is no impulsive step change in a, and second because the operation can be spread out over a longer period resulting in a lower value of a_{max}. This gives a more viable cam for high speed work.

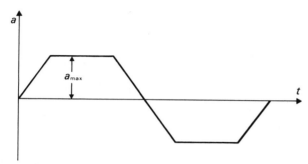

Figure 35 Trapezoidal acceleration–time graph

An even better method being adopted is to smooth out the acceleration–time curve either to a sine wave or even to a polynomial curve. The possibilities in the latter case are limitless – if they are worth it. The maximum accelerations can be made equal, that is a_a can equal a_d, and the build-up of forces can be arranged to the desired value in any fashion, at the same time maintaining *jerk*, or rate of change of acceleration, at a low and controlled level.

However, a very acceptable and simpler profile is the cycloidal cam form. A cycloid is the curve described by a point on the circumference of a circle which rolls without slipping along a straight line. The relevant curves are shown in Figure 36. The total travel S of the follower has been divided into z equal parts, where here $z = 8$. The angle of rotation of the generating circle which rolls along the s axis is β, so that when n divisions have been traversed,

$$\beta = \frac{2\pi n}{z}.$$

The radius of the generating circle is l where

$$2\pi l = S.$$

The time for the follower to move from 0 to S is T. Intermediate displacements and times are given the general symbols s and t respectively. We first want to find an equation for s in terms of t. The straight line shown is the graph of

$$s = S\frac{t}{T},$$

and represents the motion of the centre of the generating circle. What we are interested in is the motion of the point on the generating circle which at $t = 0$ coincides with the origin.

When the centre of the generating circle has moved through one division, the vertical displacement of the point we are interested in will be

$$s = S\frac{t}{T} - l\sin\beta$$

where $\beta = 45° = 2\pi/8$ rad.

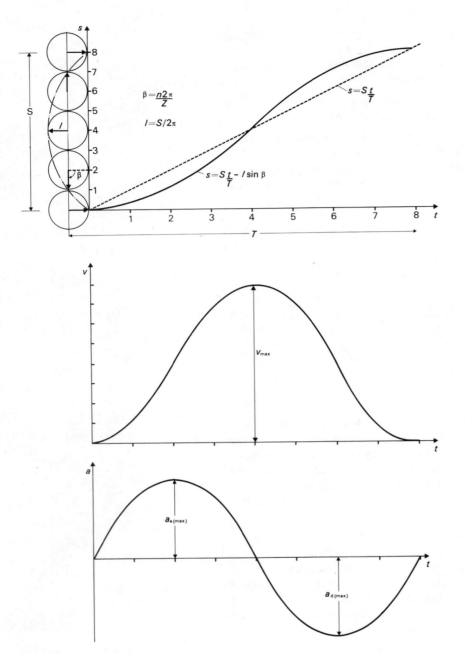

Figure 36   Displacement–time, velocity–time and acceleration–time graphs for a cycloid cam. These are not exact and are intended for illustration only

When the centre of the generating circle has moved to the second division

$$s = S\frac{t}{T} - l\sin\beta$$

where $\beta = 90° = 2\pi/4$ rad and so on.

However $l = S/2\pi$ and $\beta = 2\pi(n/z) = 2\pi(t/T)$, so that

$$s = S\left(\frac{t}{T} - \frac{1}{2\pi}\sin\frac{2\pi t}{T}\right).$$

By differentiation

$$v = \frac{S}{T}\left(1 - \cos 2\pi\,\frac{t}{T}\right)$$

and by further differentiation

$$a = \frac{2\pi S}{T^2}\sin 2\pi\,\frac{t}{T},$$

36

which enables us to construct the *v–t* diagram and the *a–t* diagram (Figure 36). We can see that the *a–t* diagram gives a smooth progression from zero to maximum acceleration, back to zero, and then on to the maximum retardation before returning to zero.

This cam form is good for high speed work. We must now see how we can apply these equations to the reality of a machine. In doing so we will again come up with problems concerning the masses of the moving parts and the physical form and dimensions of these parts.

4.2 Cam springs

In practice, the cam follower and other moving parts directly linked to the follower have mass: once they are set in motion they have a momentum which has to be dissipated before they come to rest, thereby giving rise to forces. It may be that the frictional or gravitational forces on these parts are sufficient to retain contact between the cam and follower at very low speeds of operation, but this certainly does not apply at higher speeds. The faster the follower is hurled away from the cam centre, the faster it must be returned for the next stroke.

Of course we can force contact between the two by means of a spring. But how strong a spring? Let us set up a problem – which is somewhat idealized – on an overhead camshaft for an IC engine (Figure 37).

Let us take a symmetrical cam and consider only the critical stages of the follower on the downhill side of the cam. We will assume that the centre of the cam and the centre line of movement are in line: this avoids the messy problem of major sideways forces. We will also choose the cycloidal cam form which we analysed in the last section. As the cam is symmetrical, the mathematical expression describing its shape for the 'uphill' part of the cam to its highest point of rise is the same as that for the 'downhill' return stroke.

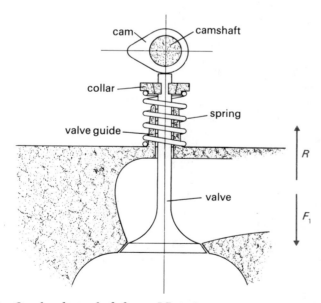

Figure 37 Overhead camshaft for an IC engine

The cam spring has to exert sufficient force to close the valve against the external load and at the same time give the valve the required acceleration. The forces on the valve during the first half of the return stroke (Figure 37) are:

R = spring force − force due to cam,

F_1 = external load.

F_2 = accelerating force on follower.

Thus

$$R - F_1 = ma = F_2$$

that is

$$R = F_1 + F_2,$$

and

$$\text{spring force} = F_1 + F_2 + \text{force due to cam}.$$

It will be easiest to work through the problem using the following data:

mass of follower $m = 0.1\,\text{kg}$;

stroke $S = 1\,\text{cm}$;

time of stroke $T = \dfrac{1}{100}\,\text{s}$.

The external load during the return stroke might have many causes but we will consider it to be due to the flow of gas through the valve and will assume that it is directly proportional to speed. Thus the external load F_1 is given by

$$F_1 = 15v.$$

We now calculate the displacement, velocity, and acceleration for the second half of the return stroke using the equations

$$s = S\left(\frac{t}{T} - \frac{1}{2\pi}\sin 2\pi \frac{t}{T}\right), \tag{10}$$

$$v = \frac{S}{T}\left(1 - \cos 2\pi \frac{t}{T}\right), \tag{11}$$

$$a = \frac{2\pi S}{T^2}\sin 2\pi \frac{t}{T}. \tag{12}$$

Taking the data given we can construct Table 5, which gives s, v and a for various values of t/T, and hence the forces F_1, F_2 and R. The velocity and acceleration are both directed towards the axis of the cam.

Table 5 Magnitudes of the displacement, velocity and acceleration

(t/T)	(s/S)	$v/\text{m s}^{-1}$	$a/\text{m s}^{-2}$	$F_1 = 15v/\text{N}$	$F_2 = ma/\text{N}$	$R = (F_1 + F_2)/\text{N}$
0.5	0.500	2.000	0.00	30.0	0.00	30.0
0.6	0.6935	1.8090	369.3	27.1	36.9	64.0
0.7	0.8534	1.3090	597.6	19.6	59.8	79.4
0.75	0.9091	1.000	628.4	15.0	62.8	77.8
0.80	0.9513	0.6910	597.6	10.4	59.8	70.2
0.85	0.9787	0.4122	508.1	6.2	50.8	57.0
0.90	0.9935	0.1910	369.3	2.9	36.9	39.8
1.00	1.000	0.00	0.00	0.00	0.00	0.00

The forces F_1, F_2 and R are plotted against s/S in Figure 38. To find the maximum force exerted by the spring, we initially assume that at $s = 0$, the spring force is zero. We then draw a straight line from $s = 0$, tangential to the curve $R = F_1 + F_2$. At the point where the tangent touches the curve, the spring force is just adequate (the follower just maintains contact with the cam) and the line indicates that the maximum spring force is 99 N. The spring stiffness k is the slope of this tangent:

$$k = \frac{99}{0.01} = 9900\,\text{N m}^{-1}.$$

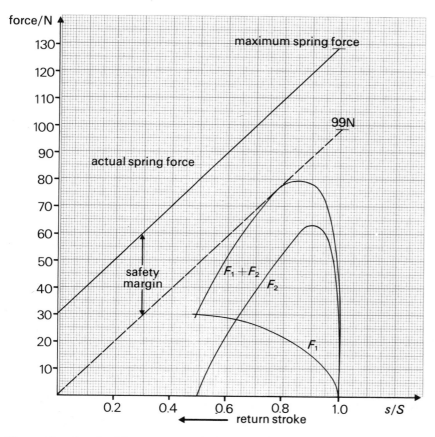

Figure 38   Graph of force against s/S for a cycloidal cam

We now add a reasonable 'safety margin' by giving the spring an initial compression (that is an initial force) at $s = 0$. This will not affect the stiffness, and so we get a second line of spring force which gives a maximum force of 129 N at $s = S$.

Notice that this calculation has not allowed for the mass of the spring, the elasticity of the components, backlash, and manufacturing tolerances. In addition, the weight of the valve has been neglected as being too small to matter.

SAQ 12

Using the same data, but assuming the s–t curve to be sinusoidal, work out the maximum values of F_1 and F_2 in the overhead cam-shaft problem.

SAQ 12

You may relate what we have done here to your experience in decoking and valve-grinding the family 'banger' and to the 'go faster' kits frequently reviewed in the motor magazines. For instance, an engine revving at 'full chat' can suffer valve 'bounce'; this is due to the follower leaving the cam, possibly due to inadequate valve springs. Figure 38 shows that if the spring stiffness line cuts the line $R = F_1 + F_2$, there will not be enough restoring force to keep cam and follower in contact. The faster the cam revolves the greater the restoring force required (equation (12) gives a squared relationship between a and the periodic time T).

High-lift cams, beloved of the 'high-speed boys', require stiffer springs; this is because, having larger values of S, they call for greater velocities and hence greater accelerations (as equations (11) and (12) show). The external load on the valve on the return stroke therefore increases (as more gas flows past the faster moving valve), resulting in greater forces on the cam surface and hence a greater rate of wear.

4.3 Development of cam profiles

From the displacement–time curve for a cam, we can construct the actual cam profile.

The simplest case is that of a point follower, the displacement–time curve for which is shown in Figure 39. The cam profile can be constructed by going through the following steps.

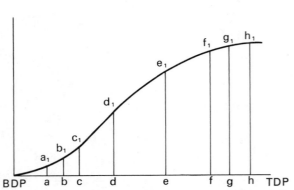

Figure 39   *Displacement–time graph of a point follower*

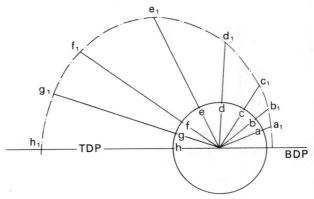

Figure 40   *Construction of the cam profile*

First divide the curve in Figure 39 into as many sections as is convenient. Second, draw a circle with radius equal to the minimum radius of the cam. Starting at a point chosen as the bottom dwell position BDP (Figure 40), mark off radial lines from the centre corresponding to the section lines $a–a_1$, $b–b_1$, etc. at angles relating to their positions on the time scale in Figure 39.

Next mark off distances $a–a_1$, $b–b_1$ etc. on the appropriate radial line. In this construction, we have effectively held the cam stationary and allowed the follower to rotate around it. We may do similar constructions with roller followers, with the provision that:

(a) each time the displacement distance of the centre of the roller is marked off, the outline of the particular follower is drawn in:

(b) when the follower is offset from the centre of the cam, the displacements are marked off along lines tangential to a circle of radius equal to the amount of offset (as in the following exercise).

Exercise

The following data refers to a cam with a sinusoidal profile: minimum radius = 30 mm; stroke = 45 mm; angle of rotation corresponding to rise = 90°; diameter of follower roller = 20 mm. Plot a 'rise' part of the profile assuming,

(a) the line of action of the follower passes through the centre of the cam;

(b) the line of action of the follower is offset 10 mm from the centre of the cam.

The construction is shown in Figure 41, which should be self-explanatory. Notice that for part (b), the pitch line is marked out by plotting 'displacement $+x_{02}$' from the tangent point between the line of action of the follower and the offset circle: x_{02} is the distance of the centre of the roller from the centre of the cam, measured in its direction of travel, when the roller is in its lowest position (similarly x_{01} for part (a)).

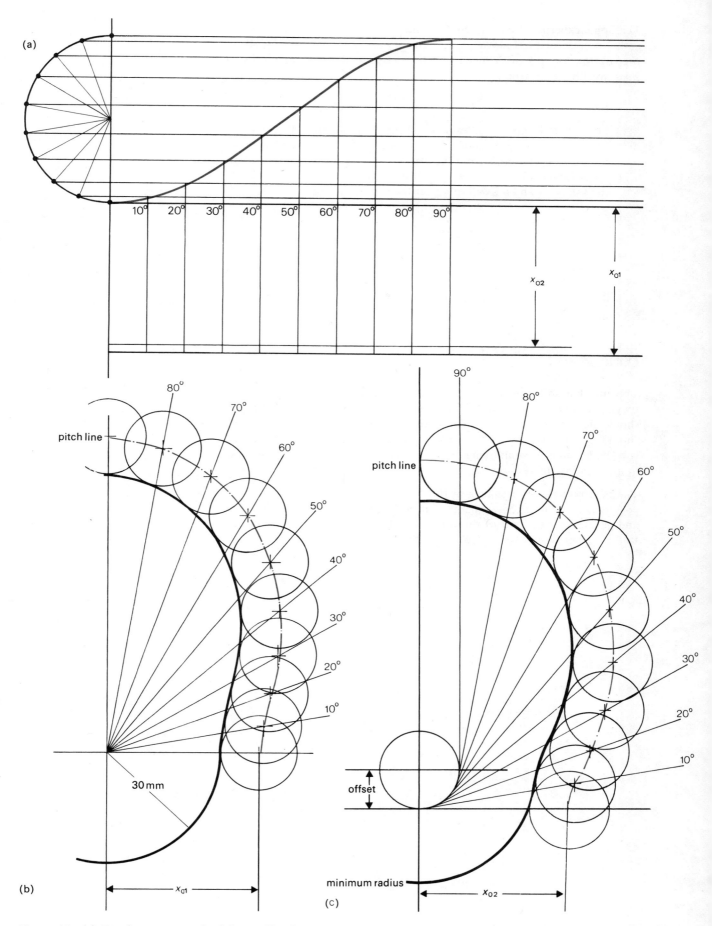

(a)

10° 20° 30° 40° 50° 60° 70° 80° 90°

x_{02} x_{01}

80° 70° 60° 50° 40° 30° 20° 10°

pitch line

30 mm

x_{01}

(b)

90° 80° 70° 60° 50° 40° 30° 20° 10°

pitch line

offset

minimum radius

x_{02}

(c)

Figure 41 (a) *Displacement graph of the profile of a sinusoidal cam;* x_{01} *and* x_{02} *are measured from the constructions in* (b) *and* (c). (b) *Displacement graph of the radial follower, with line of action through the centre of the cam.* (c) *Displacement graph of the radial follower, with line of action offset* 10 mm *from the centre of the cam*

41

The home experiment associated with this unit will help you to check the actual motion of the follower against theory. You should do SAQ 13 in preparation for this experiment, and then refer to the *Home Experiment Book* for further instructions.

SAQ 13

Plot a cam to give a follower motion corresponding to the first two columns of Table 5, and with a stroke of 5 cm. Assume a point follower offset 1.6 cm from the centre, and that the cam has a minimum radius of 6 cm. Allow 80° of cam rotation for each stroke of the follower, and a dwell of 40° at the top of the rise.

Section 5

CONCLUSION

This unit has shown us just a few practical uses of the material of earlier units. We have not, of course, been able to use all of this material, but you should now have some idea of how the various topics fit together. You should also have noticed that some of the art of engineering calculation is to make suitable simplifying assumptions and to take account of these in interpreting the results.

The next unit marks a change in the direction of this course, which now goes on to discuss energy, energy conversion, and the motion of liquids and gases. Near the end of the course we shall return to machine design studies rather similar to those begun in this unit.

APPENDIX

The parallel axes theorem for second moments of mass of a rigid body

This theorem establishes the relationship between the values obtained for the second moment of mass of any rigid body about each of two parallel axes in any direction. One of these axes must pass through the centre of mass of the body but there is no restriction on the position of the other.

Two different proofs will be given: one based on geometry, the other based on dynamics. Whichever you prefer, you should derive benefit and interest from seeing how the same result can be obtained by two different (but very simple) methods, and both provide revision of material from the course.

A1 First proof

Figure 42 shows a thin slice parallel to the plane of the paper cut from a rigid body of abitrary shape. The plane of the slice is perpendicular to two parallel axes labelled G and O which are therefore perpendicular to the plane of the diagram. The axis labelled G passes through the centre of mass of the rigid body. The shortest distance between the two axes is h.

The figure also shows a constituent particle of the slice; it has mass m. The perpendicular distances of the particle from axes G and O are r and d respectively. For the triangle formed by the lengths h, r and d, the cosine rule gives

$$d^2 = r^2 + h^2 - 2hr\cos\theta.$$

Now $\cos\theta = -\cos\phi$ since $\cos(180° - \phi) = -\cos\phi$, thus

$$d^2 = h^2 + r^2 + 2hr\cos\phi$$
$$= h^2 + r^2 + 2hx.$$

The second moment of mass of the particle about axis O is

$$md^2 = mr^2 + mh^2 + 2mhx.$$

Thus we can write the corresponding expression for the second moment of mass about O of the whole body by making each term represent all the particles: not only those in the slice considered, but those in all parallel slices of the body. This gives:

$$\Sigma md^2 = \Sigma mr^2 + h^2\Sigma m + 2h\Sigma mx. \tag{13}$$

The term h has been taken out of the summation signs because it has the same value for all particles. We can now look at each term of equation (13):

$\Sigma md^2 = I_0$ = total second moment of the body about axis O;

$\Sigma mr^2 = I_G$ = total second moment of the body about axis G;

$\Sigma m = M$ = total mass of the body;

$\Sigma mx = 0$ = first moment of mass about a plane through the centre of mass of the body, at G.

We can therefore rewrite equation (13) as follows:

$$I_0 = I_G + Mh^2.$$

This statement constitutes the parallel axes theorem.

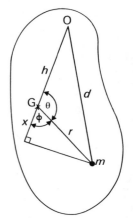

Figure 42 Thin slice from an arbitrary shape

There are three points to be noted about this theorem:

(a) Since $I = MK^2$ where K is a radius of gyration, it follows that

$$MK_0^2 = MK_G^2 + Mh^2$$

that is

$$K_0^2 = K_G^2 + h^2.$$

(b) For parallel axes in any particular direction, the second moment of mass about the axis through the centre of mass is *least*, since for this axis $h = 0$.

(c) The parallel axes theorem applies only when one of the two axes considered passes through the centre of mass of the rigid body.

A2 Second proof

Figure 43 shows a rigid body rotating about a fixed axis O perpendicular to the plane of the diagram. The body is acted on by external forces F_1, F_2 and F_3, and has an angular acceleration α as shown.

We know from our work in Units 6/7 that the moment of the external forces about the axis of rotation equals $I_0\alpha$ where I_0 is the second moment of mass about axis O. We also know that the external forces (including the force at O) can be replaced by a force acting through G, the centre of mass of the body, together with a couple in the plane of the diagram. The force equals Ma_G where M is the mass of the body and a_G is the absolute acceleration of G. The couple equals $I_G\alpha$ where I_G is the second moment of mass of the body about an axis through G parallel to the axis of rotation.

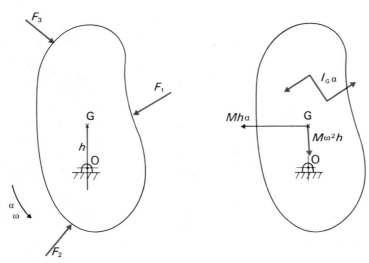

Figure 43   Rigid body rotating   Figure 44   Equivalent to Figure 43
about a fixed axis

This substitution has been made in Figure 44, which shows the force at G split into its components $M\omega^2h$ along GO and $M\alpha h$ at right angles to GO. The system of forces and moments in Figure 44 is in all respects equivalent to that in Figure 43.

Taking moments about the axis of rotation, therefore, we get

$$I_0\alpha = I_G\alpha + (Mh\alpha)h$$

(since OG $= h$ and the radial force $M\omega^2h$ has no moment about axis O). Hence

$$I_0\alpha = I_G\alpha + Mh^2\alpha,$$

giving

$$I_0 = I_G + Mh^2$$

as before.

44

ANSWERS TO SELF-ASSESSMENT QUESTIONS

SAQ 1

Distance of G_T from axis $QG_1 = (1.1 \times 0.76)\,mm$
$$= 0.84\,mm$$

Distance of G_T from axis $QO = (1.1 \times 43.6)\,mm$
$$= 47.96\,mm.$$

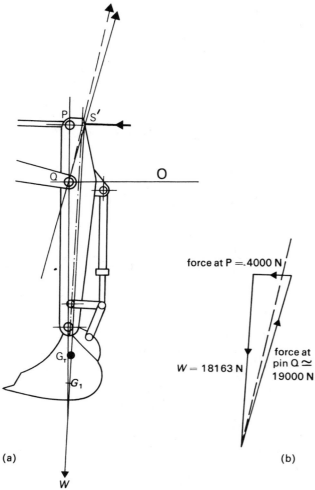

Figure 45 (a) Forces on the dipper arm: scale 1/48 full size (the original directions of the forces are indicated by dotted lines and S′ is the new position of S. (b) Vector diagram: scale 1 in $\equiv 10^4$ N

From the new force diagram (Figure 45 (b)), the new force at Q is about 19 000 N. The original value was 18 800. Hence the change in the force at Q is

$$19\,000 - 18\,800 = 200\,N.$$

Thus the percentage change is

$$\frac{200}{18\,800} \times 100 = 1.06\%$$

Hence for the evaluation of the force at Q, extreme accuracy in positioning G_T is not required.

SAQ 2

The second moment of mass of the 'spherical bucket' about an axis through G_1 (see Table 1 in Unit 6/7) is

$$\frac{2}{5} \times 1400 \times (0.048)^2 = 560 \times 23 \times 10^{-4}$$
$$= 12\,900 \times 10^{-4}$$
$$= 1.29\,kg\,m^2.$$

The corresponding value of the second moment of mass of the bucket about the pin at Q is

$$(8890 + 1.29)\,kg\,m^2.$$

The difference between this and the value given in Table 4 is negligible.

SAQ 3

The solution is shown in Figure 46.

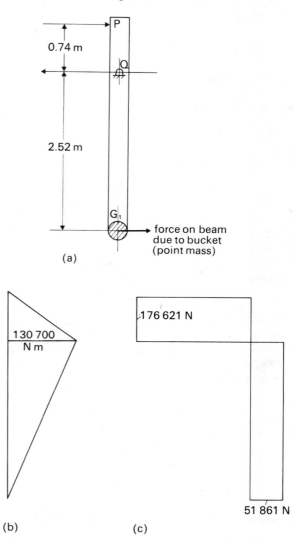

Figure 46 (a) Forces on the dipper arm. (b) Bending moment diagram. (c) Shear force diagram

The shear forces are worked out from the slopes of the bending moment diagram (remember that $dM/dx = V$). For example:

$$\frac{130\,700}{0.74} = 176\,621\,N.$$

This is the value of the force at P that would be required to accelerate the dipper arm if the assumptions we made in the calculations (for example, a point mass bucket, a light dipper arm) were realized.

SAQ 4

Referring to Figure 47 (a), the second moment of area of the section about BB is

45

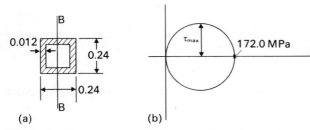

Figure 47 (*a*) *Dimensions of the section.* (*b*) *Mohr's circle*

$$\frac{(0.24)^4}{12} - \frac{(0.24-0.024)^4}{12}$$

$$= \frac{1}{12}[0.00332-(0.216)^4]$$

$$= \frac{1}{12}[0.00332-0.00218] = \frac{0.00114}{12}$$

$$= 0.000095 \, \text{m}^4.$$

Thus the maximum bending stress

$$\sigma_{max} = \frac{130\,700}{0.000095} \times \frac{0.24}{2}$$

$$= \frac{13\,070}{9.5} \times 0.12 \times 10^6$$

$$= 165.1 \times 10^6 \, \text{Pa}$$

$$= 165.1 \, \text{MPa}.$$

From Mohr's circle taking $\sigma_1 = 165$ MPa and $\sigma_3 = 0$ (Figure 47(b)) the maximum shear stress is

$$\tau_{max} = 82.5 \, \text{MPa}$$

at 45° to σ_{max}.

SAQ 5

The greatest tractive force exerted is $80\,\text{kN} = F$. Thus from Figure 48, resolving forces up the slope,

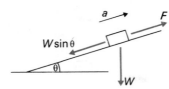

Figure 48 *Forces when the excavator is on the incline*

$$F - W\sin\theta = \frac{W}{g}a,$$

$$a = \frac{g}{W}(F - W\sin\theta) = \frac{F}{M} - g\sin\theta.$$

Since $\sin\theta = \frac{1}{5}$,

$$a = \frac{80 \times 10^3}{12 \times 10^3} - (9.81 \times \tfrac{1}{5})$$

$$= 6.66 - 1.96 = 4.70 \, \text{m s}^{-2}.$$

The instantaneous power output of the engine is

$$80 \times 10^3 \times 1.92 \times \frac{1000}{3600} = 42.7 \times 10^3 \, \text{W}$$

$$= 42.7 \, \text{kW}.$$

SAQ 6

The gas force is given by

$$F_G = \frac{\pi}{4} \times (0.098)^2 \times 200 \times 10^4$$

$$= 0.00960 \times 10^6 \times 2 \times \frac{\pi}{4} = 15 \times 10^3 \, \text{N}.$$

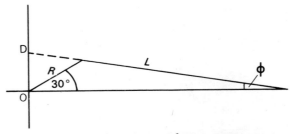

Figure 49 *Piston–crank mechanism,* $\theta = 30°$, *scale* 1 in. $\equiv$ 0.1 m

By measurement from Figure 49, OD is 0.041 m. Alternatively, by calculation:

$$r\sin\theta = l\sin\phi$$

therefore

$$\sin\phi = \frac{r}{l}\sin\theta = \frac{0.064}{0.22} \times \frac{1}{2}$$

$$= 0.145.$$

$$\phi = 8° \, 20'.$$

Thus

$$\text{OD} = (r\cos\theta + l\cos\phi)\tan\phi$$
$$= [(0.064 \times 0.866) + (0.22 \times 0.99)] \times 0.147$$
$$= (0.0554 + 0.218) \times 0.147 = 0.273 \times 0.147$$
$$= 0.0402 \, \text{m}.$$

Thus the torque at the instant considered is

$$15 \times 10^3 \times 0.04 = 6 \times 10^2 \, \text{N m}.$$

The angular velocity is given by

$$\omega = \frac{2400 \times 2\pi}{60} = 251 \, \text{rad s}^{-1},$$

thus the instantaneous power output of the engine is

$$6 \times 251 \times 10^2 = 151 \times 10^3 \, \text{W}$$

$$= 151 \, \text{kW}.$$

SAQ 7

As before, $\phi = 8° 20'$, but now we have by calculation:

$$\text{OD} = (l\cos\phi - r\cos\theta)\tan\phi$$
$$= (0.218 - 0.0554) \times 0.147$$
$$= 0.1626 \times 0.147 = 0.024 \, \text{m},$$

and from Figure 50, OD scales to 0.024 m.

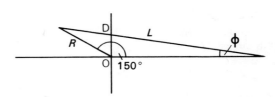

Figure 50 *Piston–crank mechanism,* $\theta = 150°$, *scale* 1 in $\equiv$ 0.1 m

Thus the torque at the instant considered is

$$2 \times 10^3 \times 0.024 = 48 \, \text{N m}$$

and the instantaneous power is

$$48 \times 251 = 12\,048 \, \text{W}$$
$$= 12.05 \, \text{kW}.$$

SAQ 8

(a) $\omega = \dfrac{2400 \times 2\pi}{60} = 251 \, \text{rad s}^{-1}$ as before.

(b) The piston acceleration is

$$\omega^2 r \left(\cos\theta + \frac{r}{l} \cos 2\theta \right)$$
$$= (251)^2 \times 0.064 \left(\cos 30° + \frac{0.064}{0.22} \cos 60° \right)$$
$$= 6.3 \times 10^4 \times 0.064(0.866 + 0.145)$$
$$= 6.3 \times 10^4 \times 0.064 \times 1.011$$
$$= 0.4076 \times 10^4$$
$$= 4076 \, \text{m s}^{-2}.$$

(c) The accelerating force F_1 is

$$2 \times 4076 = 8152 \, \text{N}.$$

(Compare F_1 with F_G and the weight of the connecting-rod.)

SAQ 9

The mass of the reciprocating parts

$$M_{\text{rec}} = 2.733 \, \text{kg},$$

and when $\theta = 30°$ we have, as before:

$$a = 4076 \, \text{m s}^{-2},$$
$$F_G = 15 \times 10^3 \, \text{N},$$
$$\text{OD} = 0.04 \, \text{m}.$$

Thus the turning moment allowing for the mass of the moving parts is

$$\text{OD} \times [F_G - M_{\text{rec}} \times a] = 0.04[15 \times 10^3 - 2.733 \times 4076]$$
$$= 0.04[15 \times 10^3 - 11.14 \times 10^3]$$
$$= 3.86 \times 10^3 \times 0.04$$
$$= 0.152 \times 10^3$$
$$= 154 \, \text{N m}.$$

SAQ 10

The greatest bending moment (at the centre of the crank pin) is

$$\tfrac{1}{2} \times Q \times \frac{0.228}{2} = 0.114 \times 20 \times 10^3$$
$$= 2280 \, \text{N m}.$$

The second moment of area is

$$I_A = \frac{\pi}{64} \times (0.075)^4 = \frac{\pi}{64} \times 0.000\,032 = 10^{-6} \times 1.57 \, \text{m}^4,$$

and since

$$\frac{\sigma}{R} = \frac{M}{I_A},$$

the magnitude of the greatest bending stress is

$$\sigma = \frac{2280}{1.57 \times 10^{-6}} \times \frac{0.075}{2}$$
$$= 10^6 \times 54.46 \, \text{N m}^{-2}$$
$$= 54.5 \, \text{MPa}.$$

This will occur at the maximum radius of the crank pin, at top and bottom in Figure 25(b).

For a stress concentration factor of three, the direct stress will be

$$54.5 \times 3 = 163.5 \, \text{MPa}$$

so that the maximum shear stress is 81.75 MPa (taking $\sigma_3 = 0$) which is still acceptable.

SAQ 11

(a) The component of Q perpendicular to the crank radius is

$$Q \cos\psi = \frac{1200}{0.064} = 18\,750 \, \text{N}.$$

(b) The component of Q along the crank radius is

$$Q \sin\psi = Q \cos\psi \times \tan\psi$$
$$= 18\,750 \tan\psi.$$

Now

$$\psi = 90° - \theta - \phi$$

where

$$\theta = 30°,$$

and

$$R \sin\theta = L \sin\phi,$$
$$\sin\phi = \frac{R}{L}\sin\theta = 0.29 \times 0.5 = 0.145,$$
$$\phi = 8.38°.$$

Thus

$$\psi = 90 - 30 - 8.38 = 51.62°,$$
$$\tan\psi = 1.262.$$

Therefore the required component

$$Q \sin\psi = 18\,750 \times 1.262 = 23\,681 \, \text{N}.$$

(c) The shear stress due to tension σ is given by

$$\frac{C}{J_A} = \frac{\sigma}{R}$$

where

$$J_A = \frac{\pi}{32} \times (0.083)^4 = \frac{0.000\,047}{32}\pi$$
$$= 10^{-6} \times 4.61 \, \text{m}^4.$$

Thus the greatest torsional shear stress is

$$\frac{1200}{4.61 \times 10^{-6}} \times \frac{0.084}{2} = 10^6 \times 10.9 \, \text{N m}^{-2}$$
$$= 10.9 \, \text{MPa}.$$

SAQ 12

The displacement-time graph is drawn in Figure 51. From this, the periodic time of one complete cycle is 0.02 s, so that

$$\frac{2\pi}{\omega} = 0.02,$$
$$\omega = \frac{2\pi}{0.02} = 100\pi.$$

Thus the equations for displacement, velocity, and acceleration are

$$s = 0.005 \sin 100\pi t,$$
$$\dot{s} = 0.5\pi \cos 100\pi t,$$
$$\ddot{s} = -50\pi^2 \sin 100\pi t.$$

The maximum acceleration is therefore $50\pi^2\,\mathrm{m\,s^{-2}}$, giving the maximum accelerating force:

$$F_2 = 0.1 \times 50\pi^2 = 5\pi^2 = 49.4\,\mathrm{N}.$$

The maximum velocity is $0.5\pi\,\mathrm{m\,s^{-1}}$, giving the maximum external load:

$$F_1 = 7.5\pi = 23.6\,\mathrm{N}.$$

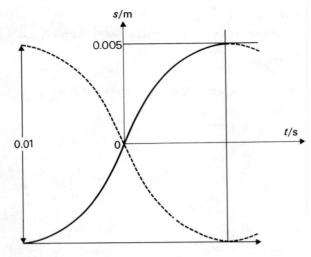

Figure 51   Displacement–time graph

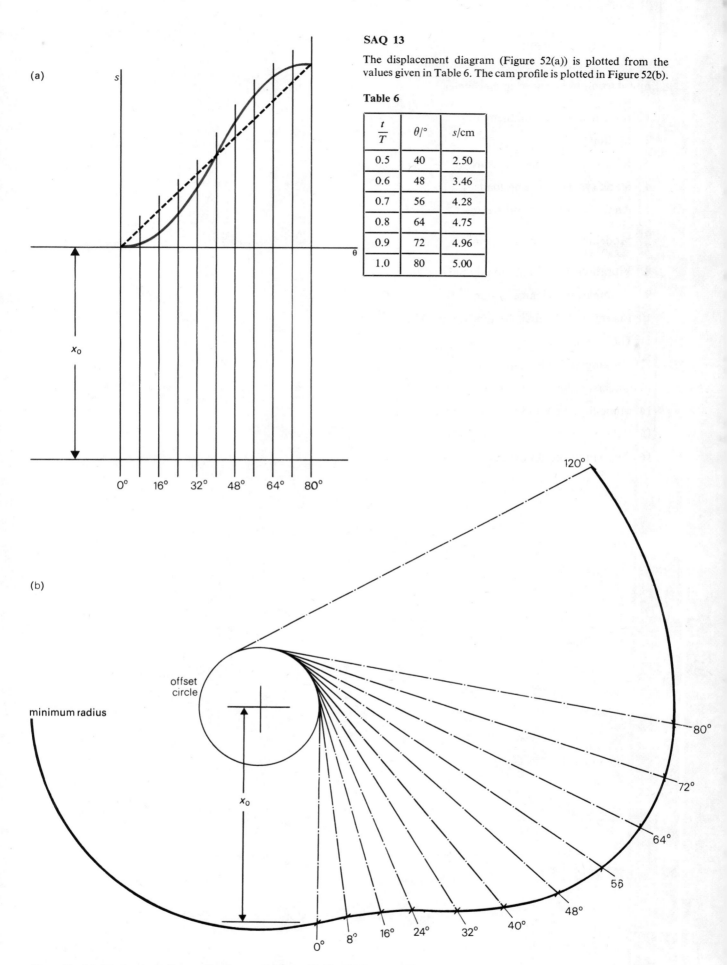

SAQ 13

The displacement diagram (Figure 52(a)) is plotted from the values given in Table 6. The cam profile is plotted in Figure 52(b).

Table 6

$\dfrac{t}{T}$	$\theta/°$	s/cm
0.5	40	2.50
0.6	48	3.46
0.7	56	4.28
0.8	64	4.75
0.9	72	4.96
1.0	80	5.00

0°   16°   32°   48°   64°   80°

x_0

(b)

offset circle

minimum radius

x_0

120°

80°

72°

64°

56°

48°

40°

32°

24°

16°

8°

0°

Figure 52   (a) Displacement diagram for the cam follower. (b) The final cam profile

49

Introduction to engineering mechanics

1 Introduction and mathematics

2 Motion

3 Kinematics and mechanisms

4 Static systems bearing loads

5 Analysing stresses and strains

6⎫
7⎭ Modelling dynamical systems

8 Vibrations and dimensional analysis

9 Examples in machine design

10 Energy and the first thermodynamic law

11 The second law

12 Looking at fluids in motion

13 Understanding fluid effects

14 Modelling fluid and thermodynamic systems

15 Case study: reciprocating engines

16 Case study: fluid flow machines and systems